TWENTY-FIRST CENTURY
SUBMARINES
AND WARSHIPS

TWENTY-FIRST CENTURY SUBMARINES AND WARSHIPS

General Editor: Peter Darman

This edition published in 2004 by Grange Books
Grange Books plc
The Grange
1–6 Kingsnorth Estate
Hoo
Near Rochester
Kent ME3 9ND
www.grangebooks.co.uk

© 2004 The Brown Reference Group plc

All rights reserved. This book is protected by copyright. No part of it may be reproduced, stored in a retrieval system, or transmitted in any form or by any means, without the prior permission in writing of the Publisher, nor be otherwise circulated in any form of binding or cover than that in which it is published and without a similar condition including this condition being imposed on the subsequent publisher.

ISBN 1-84013-678-2

Printed in China

Editorial and design:
The Brown Reference Group plc
8 Chapel Place
Rivington Street
London
EC2A 3DQ
UK
www.brownreference.com

TWENTY-FIRST CENTURY
SUBMARINES

CONTENTS

ARGENTINA	8
AUSTRALIA	9
BRAZIL	10
CHINA	11
DENMARK	14
FRANCE	16
GERMANY	22
GREECE	24
INDIA	25
ITALY	27
JAPAN	29
NETHERLANDS	32
NORTH KOREA	34
NORWAY	37
PAKISTAN	38
RUSSIA	40
SOUTH KOREA	62
SPAIN	63
SWEDEN	66
TAIWAN	70
TURKEY	71
UNITED KINGDOM	73
UNITED STATES	80

TWENTY-FIRST CENTURY WARSHIPS

CONTENTS

AUSTRALIA	99
BRAZIL	100
CANADA	101
CHINA	103
FRANCE	104
GERMANY	115
GREAT BRITAIN	118
INDIA	135
ITALY	136
NETHERLANDS	145
RUSSIA	146
SPAIN	150
UNITED STATES	152
INDEX	189

SAN JUAN

The *San Juan* is one of two Santa Cruz class attack submarines in service with the Argentine Navy. The TR1700 design, upon which the Santa Cruz class is based, comes from the Thyssen Nordseewerke shipbuilders. The original contract was for six boats to be delivered to Argentina. The first two were delivered complete and the subsequent vessels were to have materials supplied so that they could be built indigenously in Argentina's dockyards, thereby creating valuable jobs in an economy that needs such boosts desperately. However, all has not gone according to plan for the *Armada Argentina*. Two of the additional boats being built in Argentina, it is almost certain, will not be completed. Furthermore, the equipment procured for the fifth and sixth units of the class is being used for spares since Argentina cannot afford to buy spares direct from the contractor, Thyssen Nordseewerke. Without doubt, the fragile nature of Argentina's economic situation coupled with the internal political turmoil, has directly led to the drastic cutbacks being seen in its armed forces. With the future still looking uncertain for the recovery of Argentina's precarious circumstances, the two Santa Cruz submarines could well be the only attack submarines available to the Argentine Navy for some time to come, and may well serve long beyond their original service lives.

SPECIFICATIONS

Builder:	Thyssen Nordseewerke
Class:	Santa Cruz (TR1700)
Number:	S-42
Mission:	attack submarine
Length:	65.9m (217ft)
Beam:	7.3m (24ft)
Displacement:	2264 tonnes (2300 tons)
Speed:	25 knots
Operating Depth:	250m (825ft)
Maximum Depth:	300m (990ft)
Crew:	50
Nuclear Weapons:	none
Conventional Weapons:	533mm (21in) torpedo, mines
Sonar:	Krupp-Atlas CSU-83
Navigation:	unknown
Powerplant:	diesel-electric, 8000 shp
Date Commissioned:	1985

HMAS WALLER

The Royal Australian Navy's Collins class submarine is a long-range, multi-purpose attack submarine, and has been in service, despite some teething troubles, since 1999. The project is the most ambitious defence project ever undertaken in Australia, which is perhaps why there has been some difficulty in getting the boats into full operational service. It is capable of both short-duration coastal missions in littoral (shallow water) environments, and long-duration open-sea defensive and offensive operations. This makes the Collins class a very flexible weapons platform, and well suited to the demands of modern conflict and international peacekeeping or peace-enforcement. The Collins class has a patrol endurance of more than two months, and can spend most of this time submerged. However, as with all diesel-electric submarines, it needs some time on the surface to discharge exhaust fumes whilst recharging its batteries. With a displacement of over 3000 tonnes (3048 tons), the Collins class boats are some of the largest conventionally powered submarines in the world, on a similar scale to the Russian Kilo class. The project has not been cheap to implement, however, nor has it been without criticism, as the number of Collins class submarines to enter into Australian service totals six at present, at an estimated cost of over five billion Australian dollars (three billion US dollars).

SPECIFICATIONS

Builder:	Kockums ASC
Class:	Collins
Number:	75
Mission:	attack submarine
Length:	78m (257ft)
Beam:	8m (26ft)
Displacement:	3350 tonnes (3403 tons)
Speed:	20 knots
Operating Depth:	250m (825ft)
Maximum Depth:	300m (990ft)
Crew:	42
Nuclear Weapons:	none
Conventional Weapons:	533mm (21in) torpedo, Harpoon
Sonar:	Thomson Sintra Scylla sonar suite
Navigation:	Kelvin Hughes Type 1007
Powerplant:	diesel-electric, 5000 shp
Date Commissioned:	1999

TUPI

The Brazilian Navy's submarine force is based at Base Almirante Castro e Silva, Mocangue Island, just across the bay to the east of Rio de Janeiro. The force operates four Tupi class vessels and plans to have two Improved Tupi or Tikuna class submarines in service by the year 2005. The first-of-class submarine, *Tupi* (S30), was designed and built in Kiel, Germany, by Howaldtswerke Deutsche Werft (HDW) and commissioned into the Brazilian Navy in 1989. As is often the case with military foreign contract sales, the following vessels were built indigenously to the specification of the original. The Arsenal de Marinha naval shipyard in Rio de Janeiro has constructed three subsequent submarines over an extended period of 10 years, with the *Tamoio* commissioned in 1994, the *Timbira* commissioned in 1996 and the *Tapajo* commissioned in 1999. The *Tikuna*, an Improved Tupi class submarine, is under construction and is scheduled to commission in 2004, followed by the *Tapuia* in 2005. The submarine is equipped with quite sophisticated electronics, including two Mod 76 periscopes supplied by Kollmorgen, the Calypso III I-band navigation radar from Thales, and the submarine's hull-mounted sonar is the CSU-83/1 from STN Atlas Electronik. It is armed with eight torpedo tubes from which it can launch 533mm (21in) torpedoes, or mines.

SPECIFICATIONS

Builder:	Arsenal de Marinha
Class:	Tupi (Type 209/1400)
Number:	S-30
Mission:	attack submarine
Length:	61m (201ft)
Beam:	6.2m (20ft)
Displacement:	1440 tonnes (1463 tons)
Speed:	24 knots
Operating Depth:	270m (891ft)
Maximum Depth:	350m (1155ft)
Crew:	30
Nuclear Weapons:	none
Conventional Weapons:	533mm (21in) torpedo, mines
Sonar:	STN Atlas Electronik CSU-83/1
Navigation:	Thales Calypso
Powerplant:	diesel-electric, 2400 shp
Date Commissioned:	1989

CHANGZHENG 5

The Type 091, also known as the Han class, is China's first nuclear submarine. The first ship, number 401, was completed in 1974 but did not become fully operational until the 1980s due to the unreliability of its nuclear power plant and the lack of a suitable torpedo weapons system. It is believed that the French were heavily involved in aiding the upgrading of the Type 091, including help with a modified nuclear reactor with a much-reduced acoustic signature, the addition of a French-designed intercept sonar set, and the replacement of the original Soviet ESM equipment with an unknown French design. From the third ship (403) the submarine was lengthened by 8m (26ft) to be fitted with the YJ-82 anti-ship missile. All six ships are reportedly based at Jianggezhuang (Qingdao) Navy Base in the People's Liberation Army Navy (PLAN) North Sea Fleet, though other reports claimed that some of these ships may be temporarily assigned to the South Sea Fleet. The first two ships (401, 402) were refitted in the late 1980s, and back in service in the mid-1990s. Numbers 403 and 404 started mid-life refits in 1998 and were back in service in 2000. As with a great deal of Chinese equipment, unanswered questions hang over the Type 091, and it remains to be seen whether the Han class is a truly worthy attack submarine, able to stand against modern US boats.

SPECIFICATIONS

Builder:	Huludao Shipyard
Class:	Han (Type 091)
Number:	404
Mission:	attack submarine
Length:	98m (324ft)
Beam:	10m (33ft)
Displacement:	5500 tonnes (5588 tons)
Speed:	25 knots
Operating Depth:	unknown
Maximum Depth:	unknown
Crew:	75
Nuclear Weapons:	none
Conventional Weapons:	533mm (21in) torpedo, YJ-82
Sonar:	Trout Cheek sonar suite
Navigation:	Snoop Tray
Powerplant:	nuclear reactor, shp unknown
Date Commissioned:	1990

CHANGZHENG 6

The Type 092, also known as the Xia class, is China's first and only ballistic missile nuclear submarine (SSBN). It is derived from the Type 091 Han class, with the hull lengthened to accommodate the 12 missile tubes. The first ship, *Changzheng 6*, was launched in 1981 and became operational in 1983, though the JL-1 missile did not conduct a successful test launch until 1988 due to problems with its fire-control system. A second ship was reportedly constructed, but was later lost in an accident in 1985. The first test launch of the JL-1 submarine-launched ballistic missile (SLBM) took place in 1984. The first launch from a Type 092 was in 1985 but was unsuccessful, delaying the vessel's entrance into operational service. It was not until 1988 that a satisfactory launch took place. Currently the People's Liberation Army Navy (PLAN) has only one boat, *Changzheng 6*, which is deployed with the North Sea Fleet. The submarine has recently undergone an extensive refit with the upgrades reportedly including replacement of the original JL-1 missile with the improved JL-1A, which has an extended range of 4000km (2500 miles). Operations have been limited and the Type 092 has never sailed beyond Chinese regional waters. Despite a potential for operations in the Pacific Ocean, capabilities would be very limited against modern Western or Russian ASW capabilities.

SPECIFICATIONS

Builder:	Huludao Shipyard
Class:	Xia (Type 092)
Number:	406
Mission:	ballistic missile submarine
Length:	120m (396ft)
Beam:	10m (33ft)
Displacement:	6500 tonnes (6604 tons)
Speed:	22 knots
Operating Depth:	unknown
Maximum Depth:	unknown
Crew:	140
Nuclear Weapons:	12 x Ju Lang-1A SLBMs
Conventional Weapons:	533mm (21in) torpedo, mines
Sonar:	unknown
Navigation:	Snoop Tray
Powerplant:	nuclear reactor
Date Commissioned:	1986

YUANZHANG 67

China ordered two Kilo class diesel-electric submarines from Russia in 1996, and they were delivered in 1998. These two boats are the standard export version Kilo, Type 877EKM. China also ordered another two Improved Kilo, Type 636 version of the Kilo class submarines, which were delivered in 2000. The Kilo class is undoubtedly the most capable submarine in service with the People's Liberation Army Navy (PLAN). Once their crews are fully trained, these new diesel submarines will provide a substantial improvement in China's attack submarine capability. They will enhance China's capability to interdict commercial or naval shipping, and hence deny sea control to potentially hostile forces operating in China's coastal areas. Currently all four Kilo vessels are deployed as part of the East Sea Fleet and ought to remain as one of China's most important naval assets well into the twenty-first century. Reportedly, however, the first two Kilo submarines have engaged in only limited sea operations due to engine problems. Latest reports appear to confirm that China has recently ordered another eight Type 636 Improved Kilo class submarines from Russia, for a total of 12 boats in service by 2005. These submarines may be fitted with an Air Independent Propulsion (AIP) system and carry the Klub submerged-launch anti-ship missiles.

SPECIFICATIONS

Builder:	Huludao Shipyard
Class:	Kilo
Number:	367
Mission:	attack submarine
Length:	73m (241ft)
Beam:	9.9m (33ft)
Displacement:	3000 tons (3048 tons)
Speed:	20 knots
Operating Depth:	300m (990ft)
Maximum Depth:	400m (1320ft)
Crew:	52
Nuclear Weapons:	none
Conventional Weapons:	533mm (21in) torpedo, Klub
Sonar:	MGK-400M
Navigation:	unknown
Powerplant:	diesel-electric, 6000 shp
Date Commissioned:	1991

HMS KRONBORG

The original HMS *Nacken* was launched in 1978 as the lead submarine in the Nacken class for the Swedish Navy, a class consisting of three submarines. However, HMS *Nacken* has recently been extensively modernized, and was the test platform for the Air Independent Propulsion (AIP) system built into the new Gotland class and the refitted Sodermanland class. On the strength of its performance in these trials, and the fact that it would remain fitted with the AIP system, it was a very attractive prospect for purchase. Thus the Danish Navy stepped in with a bid to buy it, and in 2001 the *Nacken* was commissioned into service with the Danish Navy and renamed HMS *Kronborg*. The purchase of this vessel marked a significant upgrading of the Danish Navy's war-fighting capabilities. Its existing fleet of Tumleren class boats lacked the technology of the Nacken, and the Danish Navy found itself less able to cooperate with its allies. With the introduction of HMS *Kronborg*, however, the technology gap between the Danish Navy and her Scandinavian neighbours has now been plugged until the pan-Scandinavian Viking project comes into service. In addition to the improvement in capability, HMS *Kronborg* provides Danish crews with vital experience of a modern vessel, which will be invaluable when the next-generation boats enter Danish service.

SPECIFICATIONS

Builder:	Kockums
Class:	Kronborg
Number:	S-325
Mission:	attack submarine
Length:	57.9m (191ft)
Beam:	5.7m (19ft)
Displacement:	1218 tonnes (1237 tons)
Speed:	20 knots
Operating Depth:	250m (825ft)
Maximum Depth:	350m (1155ft)
Crew:	19
Nuclear Weapons:	none
Conventional Weapons:	533mm (21in) torpedo, mines
Sonar:	Atlas Electronik CSU-83 sonar
Navigation:	Terna radar
Powerplant:	diesel-electric AIP, 1500 shp
Date Commissioned:	2001

HMS SÆLEN

The three submarines of the Tumleren class are of the German Type 207 design. In 1964–66 they were built at the Rheinstahl-Nordseewerke in Germany for the Norwegian Navy as part of the 15 units of the Kobben class. In 1989–91 three units of the Kobben class were purchased and taken over by the Royal Danish Navy from Norway. Following an extensive refit they were renamed *HMS Tumleren*, *HMS Sælen* and *HMS Springeren* and replaced the four boats of the older Danish-designed and built Delfinen class. The Danish submarines are equipped and trained for an emphasis on littoral operations. This reflects the geography of Denmark and its historical and strategic position of controlling the entrance to the Baltic Sea. Since the birth of the Danish submarine force in 1909, the ability to operate in shallow and confined waters has been of prime importance. In addition, the more traditional open-water operations in the North Atlantic region are exercised routinely by participation in NATO's Standing Naval Force Atlantic. Following the end of the Cold War and NATO's new strategic concept, the tasks of the Danish armed forces have been redirected towards international operations such as peacekeeping and peace-enforcement. Such operations will typically require enhanced littoral capabilities encompassing all the different kinds of submarine warfare.

SPECIFICATIONS

Builder:	Howaldtswerke-Deutsche Werft
Class:	Tumleren
Number:	S-323
Mission:	coastal patrol
Length:	46.6m (154ft)
Beam:	4.6m (15ft)
Displacement:	524 tonnes (532 tons)
Speed:	17 knots
Operating Depth:	200m (660ft)
Maximum Depth:	270m (891ft)
Crew:	18
Nuclear Weapons:	none
Conventional Weapons:	533mm (21in) torpedo, mines
Sonar:	PSU-83
Navigation:	unknown
Powerplant:	diesel-electric, 1700 shp
Date Commissioned:	1990

AMÉTHYSTE

The French Navy's Améthyste class attack submarine is a follow-on development from the successful Rubis class SSN. It has been designed for littoral anti-submarine operations, but like most modern submarines can perform a varied number of jobs with efficiency, including anti-shipping, reconnaissance and power-projection missions. The Améthyste class incorporates many improvements on the original Rubis design to make it a more capable machine. These include a slightly longer hull to incorporate a quieter propulsion system, and a much improved electronics suite. In addition to the 533mm (21in) torpedoes, the Améthyste class can launch the ship-killing Exocet missile. The missile is guided by target range and bearing data from the submarine's tactical data system and weapons control system, which is downloaded into the Exocet's onboard computer. The missile approaches the target area in a sea-skimming fashion using inertial navigation to avoid detection, and then uses active radar homing as it nears the target. The missile approaches the target at speeds of over Mach 0.9 (1170 km/h) and the range is 50km (31.25 miles). The Exocet is armed with a 165kg (363lb) high-explosive shaped charge and is utterly deadly. Exocet missiles sank the British ships HMS *Sheffield* and *Atlantic Conveyor* during the 1982 Falklands War.

SPECIFICATIONS

Builder:	DCN International
Class:	Améthyste
Number:	S-605
Mission:	attack submarine
Length:	73.6m (241ft)
Beam:	7.6m (25ft)
Displacement:	2730 tonnes (2774 tons)
Speed:	25 knots
Operating Depth:	200m (660ft)
Maximum Depth:	300m (990ft)
Crew:	70
Nuclear Weapons:	none
Conventional Weapons:	533mm (21in) torpedo, Exocet
Sonar:	Thomson Sintra DMUX 20
Navigation:	Thomson-CSF DRUA 33
Powerplant:	nuclear reactor, 9500 shp
Date Commissioned:	1992

EMERAUDE

The French Navy has modified four of its older Rubis class attack submarines, to bring them up to a similar standard to the more modern Améthyste class SSNs. The resulting hybrid class is known as the Rubis-Améthyste. The French Navy operates its Rubis-Améthyste class submarines from the naval base in Toulon. The first batch of Rubis class submarines was initially equipped for an anti-surface role, but since becoming the Rubis-Améthyste class they have been re-equipped to the same standard as the later Améthyste submarines, giving them the capability to perform both anti-surface and anti-submarine warfare missions. This has been achieved by the addition of two extra torpedo tubes, and improved sonar, radar and navigation systems. The submarine has the capacity to carry 14 missiles and torpedoes in a mixed load. The four 533mm (21in) torpedo tubes are equipped with a pneumatic ram system for discharging torpedoes from the tubes. The Rubis-Améthyste is armed with two types of torpedo. One is the ECAN L5 Mod 3 torpedo. This is equipped with active and passive homing and has a range of 9.5km (6 miles). The torpedo has a speed of 35 knots and delivers a 150kg (330lb) warhead to a depth of 550m (1815ft). The ECAN F17 Mod 2 torpedo is a wire-guided torpedo with a range of over 20km (12.5 miles) and a depth of up to 600m (1980ft).

SPECIFICATIONS

Builder:	DCN International
Class:	Rubis-Améthyste
Number:	S-604
Mission:	attack submarine
Length:	72.1m (236ft)
Beam:	7.6m (25ft)
Displacement:	2670 tonnes (2712 tons)
Speed:	25 knots
Operating Depth:	250m (825ft)
Maximum Depth:	300m (990ft)
Crew:	70
Nuclear Weapons:	none
Conventional Weapons:	533mm (21in) torpedo, Exocet
Sonar:	Thomson Sintra DMUX 20
Navigation:	Thomson-CSF DRUA 33
Powerplant:	nuclear reactor, 9500 shp
Date Commissioned:	1988 (recommissioned 1996)

L'INDOMPTABLE

The French Navy's *L'Indomptable* vessel has been retained in service despite being slated for decommissioning in the 1990s, and is expected to remain so until 2004. This vessel carries a compliment of 130 men, of whom 15 are officers. Like most ballistic missile submarines (SSBNs) the crew is divided into two, *rouge* (red) and *bleu* (blue), who keep a continual 24-hour watch. This means that the vessel can react to emergency order immediately, rather than having to prepare. The ability to carry out emergency orders immediately is a key part of strategic deterrence, since any potential aggressor must be aware that attack will be met with an immediate response. The French Navy's SSBNs have not seen as much active service as the US or British fleets, but they continue to patrol the world's oceans protecting France's interests. Whereas the British SSBN submarine fleet takes response orders and targeting parameters from the US, the French president retains the sole right to give the order to launch France's nuclear weapons. In reality, however, France is unlikely to act alone in any given international crisis where the threat of nuclear force is present. The M4 submarine-launched ballistic missile (SLBM) that is carried by the L'Inflexible class is capable of hitting six different targets up to 4000km (2500 miles) away.

SPECIFICATIONS

Builder:	DCN International
Class:	L'Inflexible
Number:	S-613
Mission:	ballistic missile submarine
Length:	128m (420ft)
Beam:	10.5m (35ft)
Displacement:	9000 tonnes (9144 tons)
Speed:	25 knots
Operating Depth:	280m (924ft)
Maximum Depth:	350m (1155ft)
Crew:	130
Nuclear Weapons:	16 x M4 SLBM
Conventional Weapons:	533mm (21in) torpedo, Exocet
Sonar:	Thomson Sintra DSUX 21
Navigation:	Thomson-CSF DRUA 33
Powerplant:	nuclear reactor, 16,000 shp
Date Commissioned:	1976 (recommissioned 1989)

L'INFLEXIBLE

The L'inflexible class ballistic missile submarines were the result of a modernizing of the ageing Le Redoubtable class SSBNs launched in the early 1970s. The French have traditionally developed their own nuclear technology, which meant that their sea-based nuclear deterrence was lagging far behind the US and US-aided British SSBNs. When the original Le Redoubtable class was launched it was already behind the times, and thus a modernization plan brought the class up to L'Inflexible standards. When the vessel bearing the name *Le Redoubtable* was decommissioned in 1991, all remaining boats became known as L'Inflexible class, irrespective of their original pre-overhaul configuration. This vessel is part of France's Strategic Oceanic Force (FOST), but it is due to retire from active duty in the next few years as the Le Triomphant class begins to enter service. Delays with the remaining Le Triomphant class vessels means that the L'Inflexible class boats will stay in service longer than the original 2003 retirement date. The last remaining vessel of the original L'Inflexible class has been recently retro-fitted with the M45 SLBM to reflect the fact that it will be in service longer than expected. Though no match for the most modern SSBNs around the world, they are still capable vessels and maintain France's nuclear deterrence.

SPECIFICATIONS

Builder:	DCN International
Class:	L'Inflexible
Number:	S-615
Mission:	ballistic missile submarine
Length:	128m (420ft)
Beam:	10.5m (35ft)
Displacement:	9000 tonnes (9144 tons)
Speed:	25 knots
Operating Depth:	280m (924ft)
Maximum Depth:	350m (1155ft)
Crew:	130
Nuclear Weapons:	16 x M45 SLBM
Conventional Weapons:	533mm (21in) torpedo, Exocet
Sonar:	Thomson Sintra DSUX 21
Navigation:	Thomson-CSF DRUA 33
Powerplant:	nuclear reactor, 16,000 shp
Date Commissioned:	1985

LE TÉMÉRAIRE

The new Le Triomphant class of ballistic missile submarine is an integral part of France's Strategic Oceanic Force (FOST). Much like the United Kingdom, French strategic doctrine demands that at least one SSBN be on patrol at any given time, able to launch a retaliatory second-strike in the event of an attack on France itself or any of her NATO allies. Though the world is a very different place since the collapse of communism in Russia, France still sees a necessity to develop and maintain an SSBN force, as well as keeping a nuclear bombing capability. The modern Le Triomphant class takes advantage of many developments in computer technology and materials advances to keep it ahead of its predecessor, the L'Inflexible class SSBN which entered service in the 1980s. It boasts a sophisticated sensor array that gives a very extensive detection range, and has a modern steel structure composed of US equivalent HY130 steel. This gives the class the ability to dive much deeper than its predecessor, thus making it more flexible and stealthy. Le Téméraire is the second vessel in a batch of four Le Triomphant class submarines to be launched, and by 2008 all four should be in service. Under the SNLE-NG (Sous-Marins Nucléaires Lanceurs Engins-Nouvelle Génération) programme, each Le Triomphant SSBN costs 5.5 billion Euros.

SPECIFICATIONS

Builder:	DCN International
Class:	Le Triomphant
Number:	S-617
Mission:	ballistic missile submarine
Length:	138m (452ft)
Beam:	12.5m (41ft)
Displacement:	14,120 tonnes (14,345 tons)
Speed:	25 knots
Operating Depth:	300m (990ft)
Maximum Depth:	400m (1320ft)
Crew:	110
Nuclear Weapons:	16 x M45 ballistic missiles
Conventional Weapons:	533mm (21in) torpedo, Exocet
Sonar:	DMUX 80 sonar suite
Navigation:	Racal 1229 DRBN 34A
Powerplant:	nuclear reactor, 41,000 shp
Date Commissioned:	1997

LE TRIOMPHANT

The French Navy's Le Triomphant class ballistic missile submarine is the replacement for the L'Inflexible M4 class SSBNs. The submarine was designed and built at DCN's (Direction des Constructions et Armes Navales) Cherbourg shipyard, and the first-of-class submarine was launched in July 1993 and entered service in 1997. There are to be four submarines in total in this class. The *Le Triomphant* carries 16 vertically launched M45 ballistic missiles with a warhead of six multiple re-entry vehicles (MRVs). Each MRV has the explosive power equivalent to 147,638 tons (150,000 tons) of TNT, making it 10 times more powerful than the A-bomb dropped on Hiroshima. The missile has a range of 6000km (3750 miles). A new enhanced M51 ballistic missile is due to enter service in 2008, and this will carry a warhead with 12 multiple independently targetable re-entry vehicles (MIRVs), and have an increased range of 8000km (5000 miles). Aside from its ballistic missiles, *Le Triomphant* carries the Exocet surface-to-surface anti-shipping missile, as well as a variety of torpedoes fired from its four 533mm (21in) torpedo tubes. France has invested large sums of money in modernizing and maintaining its strategic nuclear deterrence, and in *Le Triomphant* it has a vessel to take their nuclear capability into the twenty-first century.

SPECIFICATIONS

Builder:	DCN International
Class:	Le Triomphant
Number:	S-616
Mission:	ballistic missile submarine
Length:	138m (452ft)
Beam:	12.5m (41ft)
Displacement:	14,120 tonnes (14,345 tons)
Speed:	25+ knots
Operating Depth:	300m (990ft)
Maximum Depth:	400m (1320ft)
Crew:	110
Nuclear Weapons:	16 x M45 ballistic missiles
Conventional Weapons:	533mm (21in) torpedo, Exocet
Sonar:	DMUX 80 sonar suite
Navigation:	Racal 1229 DRBN 34A
Powerplant:	nuclear reactor, 41,000 shp
Date Commissioned:	1997

U18

The original Type 206 class attack submarines in the German Navy were commissioned between 1973 and 1975. By the end of the 1980s, a major programme of modifications upgraded 12 boats to the Type 206A class. This modernization was geared chiefly towards the introduction of an integrated computer-based combat and sonar system, and was completed in 1992. The Type 206A benefits from excellent manoeuvrability and depth-keeping capabilities, which enables it to operate submerged in water depths of just 18m (60ft). This is a great advantage in littoral operations where the seafloor often hinders larger submarines. Due to its small size, the Type 206A is quite an elusive target, both in terms of initial detection, and then tracking once found. The complement of 25 consists of 6 officers, 6 chief petty officers and 13 petty officers. The weapons control system is, among other features, capable of tracking several targets fully automatically and can handle a maximum of three wire-guided torpedoes simultaneously. When approaching the targets, the torpedoes will gain contact and lock their homing heads onto the target, but do not attack unless ordered to do so by the operator of the control system. Fitted with eight sophisticated torpedo tubes for use against surface and sub-surface vessels, the 206A has considerable firepower for its relatively diminutive size.

SPECIFICATIONS

Builder:	Howaldtswerke-Deutsche Werft
Class:	Type 206A
Number:	S-197
Mission:	attack submarine
Length:	48.6m (159ft)
Beam:	4.7m (15ft)
Displacement:	520 tonnes (528 tons)
Speed:	17 knots
Operating Depth:	180m (594ft)
Maximum Depth:	300m (990ft)
Crew:	25
Nuclear Weapons:	none
Conventional Weapons:	533mm (21in) torpedo, mines
Sonar:	Thomson Sintra DUUX 2
Navigation:	Thomson CSF Calypso II
Powerplant:	diesel-electric, 2300 shp
Date Commissioned:	1975

U31

The Type 212A class attack submarine is the latest in a long and illustrious line of German diesel-electric submarines from the HDW shipyard. The German ship-building industry is the largest exporter of submarines in the world, and the Type 212A has already been ordered by a number of different countries including Italy and Norway. The new Type 212A was designed for the German Navy, amongst others, with the faults of the Type 206 class vessels in mind, and the improvements in the new class attempt to address these. The modifications include a greater long-range reconnaissance and detection capability, as well as the introduction of a satellite communications system for a fast and secure means of exchanging information. Other modifications include a further improvement on the stealthiness of the submarine, faster and longer-range torpedoes, a high degree of automation, an improvement of living conditions onboard, and an improved environmental protection system. The most important technology to be included in the new class, however, is the Air Independent Propulsion (AIP) system. In this system, oxygen and hydrogen are combined in a catalytic way to release electrical energy (to power the vessel) and the waste products. This remarkable technology makes the Type 212A class exceptionally quiet, robust and thus more deadly.

SPECIFICATIONS

Builder:	Howaldtswerke-Deutsche Werft
Class:	Type 212A
Number:	S-181
Mission:	attack submarine
Length:	56m (185ft)
Beam:	7m (23ft)
Displacement:	1830 tonnes (1828 tons)
Speed:	20 knots
Operating Depth:	300m (990ft)
Maximum Depth:	400m (1320ft)
Crew:	27
Nuclear Weapons:	none
Conventional Weapons:	533mm (21in) torpedo, mines
Sonar:	DBQS-40 sonar suite
Navigation:	Kelvin Hughes Type 1007
Powerplant:	diesel-electric AIP, 4243 shp
Date Commissioned:	2003

PONTOS

The Glavkos class patrol/attack submarine is essentially a Greek version of the popular German-built Type 209 class. The Hellenic Navy was the first in the world to order this class of submarine, and the first vessel entered service in 1971. There are eight individual boats, in two variations in service with the Hellenic Navy. The first four boats of the class are Type 1100 and the last four are Type 1200, of which the *Pontos* is one. Despite the different designations, they are considered to be the same class. The Glavkos class is a relatively simple design, and is only moderately automated, making it somewhat outdated in the modern era. The Greek Government has sought to remedy this technological shortcoming by ordering the new Type 214 submarine from the German HDW shipbuilding company. The first four boats of the class have been overhauled in the last few years, giving them the capability to launch the Sub Harpoon anti-shipping missile. This has greatly increased their war-fighting ability. In addition to the Sub Harpoon surface-to-surface missile, the Glavkos class is equipped with eight torpedo tubes capable of firing 533m (21in) torpedoes. The Glavkos class is due to remain in service until the Hellenic Navy has completed its procurement of the enhanced Type 214 class submarines, also from the same German shipbuilders.

SPECIFICATIONS

Builder:	Howaldtswerke-Deutsche Werft
Class:	Glavkos (Type 209/1200)
Number:	S-119
Mission:	patrol/attack submarine
Length:	55.9m (184ft)
Beam:	6.3m (21ft)
Displacement:	1285 tonnes (1306 tons)
Speed:	22 knots
Operating Depth:	300m (990ft)
Maximum Depth:	400m (1325ft)
Crew:	31
Nuclear Weapons:	none
Conventional Weapons:	533mm (21in) torpedo, Harpoon
Sonar:	Krupp 3-4 active/passive sonar
Navigation:	Calypso II radar
Powerplant:	diesel-electric, 5000 shp
Date Commissioned:	1980

INS SHANKUL

After several years of discussion with Howaldtswerke-Deutsche Werft (HDW), the Indian Navy came to an agreement in December 1981 to purchase a number of the popular Type 209/1500 attack submarines for its fleet. The agreement clarified that the building in West Germany of two HDW 209 class vessels would be followed by a supply of packages for building two more at Mazagon Dockyards Ltd. In addition, HDW agreed to supply training to various groups of specialists for the design and construction of the last two submarines in the batch. The first two submarines sailed for India in February 1987, having been commissioned into Indian service the year before. In 1984 it was announced that two more submarines would be built at the Mazagon Dockyards Ltd (MDL) in Mumbai, but this was overtaken by events during the Kashmiri insurgency of 1987–88 and the agreement with HDW was terminated at just four submarines, rather than the original six. This was reconsidered in 1992 and again in 1997, but no orders have yet been placed. The Shishumar class began an extensive refit during 2000, which is thought to include improvements to the electronics suite, plus other minor improvements. The class can expect to be in service until the next decade of the twenty-first century, depending on certain economic and political factors.

SPECIFICATIONS

Builder:	Mazagon Dockyards Ltd/HDW
Class:	Shishumar (Type 209/1500)
Number:	S-47
Mission:	attack submarine
Length:	64.4m (213ft)
Beam:	6.5m (22ft)
Displacement:	1850 tonnes (1880 tons)
Speed:	22 knots
Operating Depth:	260m (863ft)
Maximum Depth:	350m (1155ft)
Crew:	40
Nuclear Weapons:	none
Conventional Weapons:	533mm (21in) torpedo, mines
Sonar:	Electronik CSU-83 sonar suite
Navigation:	Thomson-CSF Calypso
Powerplant:	diesel-electric, 4600 shp
Date Commissioned:	1994

INS SINDHUVIJAY

A total of 10 diesel-powered Kilo project 877EKM submarines, known in India as the Sindhugosh class, have been built under a contract between Russian shipbuilder Rosvooruzhenie and the Indian Defence Ministry. Kilo class submarines have been nicknamed "Black Hole" by NATO for their silent operation whilst on operations. In January 1997 two Improved Kilo class boats, also known as Kilo Type 636, were ordered by the Indian Navy and the first, INS *Sindhurakshak*, was commissioned in December 1997 at St Petersburg, Russia. This submarine was a spare Type 877EKM hull built for the Russian Navy, but was never purchased. The second, INS *Sindhushastra*, was commissioned in July 2000. The remaining Sindhugosh class vessels are the standard export versions of the 877EKM Kilo. The Sindhugosh class is fitted with six 533mm (21in) torpedo tubes, carrying 18 heavyweight torpedoes (6 in the tubes and 12 on the racks). It uses an automatic rapid loader which is remotely controlled from the main control panel or by the controls in the launch station. Two of the tubes can fire wire-guided torpedoes, whilst the other four have automatic reloading. Torpedo types include the Type 53-65 passive wake-homing torpedo, and the TEST 71/76 anti-submarine active and passive homing torpedo. It can also launch mines as well as shoulder-launched SAMs.

SPECIFICATIONS

Builder:	Admiralty Shipyard (Sudomekh)
Class:	Sindhugosh (Kilo 877EKM)
Number:	S-62
Mission:	attack submarine
Length:	73m (241ft)
Beam:	9.9m (33ft)
Displacement:	3076 tonnes (3125 tons)
Speed:	17 knots
Operating Depth:	250m (825ft)
Maximum Depth:	330m (1089ft)
Crew:	53
Nuclear Weapons:	none
Conventional Weapons:	533mm (21in) torpedo, SSM
Sonar:	MGK-400 Shark Teeth
Navigation:	MG-519 Snoop Tray
Powerplant:	diesel-electric, 6800 shp
Date Commissioned:	1991

PRIMO LONGOBARDO

The Primo Longobardo class is a fourth-generation batch submarine from the original Sauro class coastal patrol vessel. The class has been modified and adapted over the years, with this class being the most modern of the fleet. Built in Italy in the shipyards of Monfalcone, the Sauro class replaced the submarines of the indigenous Toti class and the American boats of the Romei class. The *Primo Longobardo* diesel-electric attack submarine maintains the same basic design features of the original boats which entered service in 1980, however they have been upgraded technologically. A programme of modernization that started late in 1999 totally reworked the electro-acoustical sensor arrays and the command-and-control suite. The former electro-acoustical system (IPD-70S), the command-and-control system (SACTI MM/BN-716) and the launch and wire-guidance control (FCD Mk 2) system have been removed, and replaced by the ISUS 90-20, an integrated system that can more efficiently perform all the functions formerly managed by the three separate systems. The improvement in performance that the ISUS 90-20 gives also required the sound-dampening technology on the boat to be improved in order to take full advantage of the new suite's capabilities. The ISUS can guide four torpedoes at the same time against four different targets, whilst also managing mines and countermeasures.

SPECIFICATIONS

Builder:	*Italcantieri, Monfalcone*
Class:	*Primo Longobardo*
Number:	*S-524*
Mission:	*coastal patrol submarine*
Length:	*66.3m (219ft)*
Beam:	*6.8m (22ft)*
Displacement:	*1862 tonnes (1891 tons)*
Speed:	*19 knots*
Operating Depth:	*200m (660ft)*
Maximum Depth:	*300m (990ft)*
Crew:	*50*
Nuclear Weapons:	*none*
Conventional Weapons:	*533mm (21in) torpedo, mines*
Sonar:	*ISUS 90-20 sensor suite*
Navigation:	*unknown*
Powerplant:	*diesel-electric, 3650 shp*
Date Commissioned:	*1993*

SALVATORE TODARO

The latest addition to the Italian submarine fleet is the Salvatore Todaro class diesel-electric attack submarine. This class is based on the German designed Type 212A, and the Italian version differs little from the German Navy's. The introduction of the Salvatore Todaro class into the Italian Navy will be a significant step forward for the fleet. The Type 212A is a state-of-the-art diesel submarine, and its capabilities far outreach the Sauro generation of boats. The Salvatore Todaro class will not replace the Sauro generation; indeed, some of the technology from the Type 212A boats will be added to the newest Primo Longobardo class vessels, thus upgrading their capabilities. The Italian-built Type 212A vessels will be constructed by the Fincantieri company at the Monfalcone shipyards. The design is different from the German Type 212A in the life-saving systems on board. The German vessel only has provision for individual escape, whilst the Italian boats have both individual and collective life-saving features. Other than this difference, the two designs are virtually the same and share the same characteristics and capabilities. The areas in which the Salvatore Todaro class will operate are of course different from the German Navy vessels, focusing on the Mediterranean rather than the North Sea and Atlantic, protecting Italy's interests in this theatre.

SPECIFICATIONS

Builder:	Fincantieri/HDW
Class:	Type 212A
Number:	S-526
Mission:	attack submarine
Length:	56m (185ft)
Beam:	7m (23ft)
Displacement:	1830 tonnes (1828ft)
Speed:	20 knots
Operating Depth:	300m (990ft)
Maximum Depth:	400m (1320ft)
Crew:	24
Nuclear Weapons:	none
Conventional Weapons:	533mm (21in) torpedo, Harpoon
Sonar:	Electronik DBQS-40 sonar suite
Navigation:	Kelvin Hughes Type 1007
Powerplant:	diesel-electric AIP, 4243 shp
Date Commissioned:	2005

FUYUSHIO

The Harushio class attack submarine is an indigenously designed and built vessel, and uses a tear-drop type hull, unlike its successor Oyashio class. The design is technically derived from the previous Yushio class and as such there is no major advancement, but instead a few improvements including better silent-running characteristics, a reduction in noise emissions and improvements in underwater manoeuvrability. As is the case with most countries' submarine forces, many exact characteristics of the submarines of the Japanese Maritime Self Defense Force (JMSDF) are not published, with special secrecy concerning maximum depth capabilities. The Harushio class uses the NS110 high-strength steel in portions of the pressure-resistant hull, and the operating depth is presumed to be 300m (990ft) or more, some sources suggesting it to be 500m (1650ft), though this cannot be confirmed. The Harushio class is equipped with the ZQQ sonar and TASS bow sonar. The torpedo system uses the indigenously designed and built Type 89 torpedoes. The last unit of this class, the *Asashio*, was redesignated as a training submarine (the TSS 3601), and was modified for this role. It was decided in the JMSDF to use a modern vessel as the training platform, so as to model accurately the nature of modern submarine warfare and seamanship.

SPECIFICATIONS

Builder:	Kawasaki/Mitsubishi
Class:	Harushio
Number:	SS-588
Mission:	attack submarine
Length:	77m (254ft)
Beam:	10m (33ft)
Displacement:	2750 tonnes (2794 tons)
Speed:	20 knots
Operating Depth:	300m (990ft)
Maximum Depth:	400m (1320ft)
Crew:	75
Nuclear Weapons:	none
Conventional Weapons:	533mm (21in) torpedo, Harpoon
Sonar:	Hughes/Oki ZQQ-5B sonar suite
Navigation:	JRC ZPS 6
Powerplant:	diesel-electric, 7200 shp
Date Commissioned:	1995

MAKISHIO

The new Oyashio class attack submarine is an indigenously designed and built vessel that differs from previous submarines of the Japanese Maritime Self Defense Force (JMSDF). Like most submarines, a traditional tear-drop shape hull was the previously preferred design, but the latest and most powerful submarine in the JMSDF inventory has been designed with a "leaf coil" hull form. It incorporates not only a different hull form, but it employs a double-hull system, much like Russian submarines. The design of the Oyashio was built around the arrangement of the sensors first and foremost, after which the entire layout was then built. The traditional method of arrangement placed the most powerful sensors in the bow, but the Oyashio class uses its entire hull as part of the sensor array. In addition, the combat intelligence processing system of the new vessel also improves its attacking capabilities. The position of the torpedo tubes also differs from other similar submarines, with the bow torpedo tubes placement influenced by the arrangement of the sensors. The stealthiness of the hull has been improved by installing anechoic rubber tiles on the hull. High levels of automation throughout the vessel has allowed the number of crew needed to be around 75, thus saving on manpower and the logistics required to support a larger crew.

SPECIFICATIONS

Builder:	Mitsubishi/Kawasaki
Class:	Oyashio
Number:	SS-593
Mission:	attack/patrol submarine
Length:	81.7m (268ft)
Beam:	8.9m (29ft)
Displacement:	3600 tonnes (3657 tons)
Speed:	20 knots
Operating Depth:	300m (990ft)
Maximum Depth:	400m (1320ft)
Crew:	75
Nuclear Weapons:	none
Conventional Weapons:	533mm (21in) torpedo, Harpoon
Sonar:	ZQQ-5B hull/flank arrays
Navigation:	I-band radar system
Powerplant:	diesel-electric, 7700 shp
Date Commissioned:	2001

SACHISHIO

The Yushio class attack submarine represented an enlargement and improvement over the Uzushio class vessels that had preceded it when it was launched in the mid-1970s. A total of 10 were built starting in 1975, and it was the largest class of submarines built for the Japanese Maritime Self Defense Force (JMSDF) at the time. Although it was launched as a second-generation vessel, because the peripheral technology progressed massively over the decade during which they were under construction, the earliest unit and the latter units were almost different types of ships, with the later submarines technologically far superior. Gradual improvements were made over the construction period, and from the fifth boat onward these improvements included the ability to launch the Sub Harpoon missile, and in later units the control navigation system was improved. Type NS80 steel was adopted for the pressure-resistant boat hull, allowing for operations to be carried out safely at depths of approximately 450m (1485ft), though this is unconfirmed officially. Other improvements over its predecessor included the adoption of a seven-bladed propeller, which gave the submarine a reduction in noise emissions. The increase of cruising power, periscope depth range and the large-sized computer conversion of the direction device resulted in a greater war-fighting ability.

SPECIFICATIONS

Builder:	Kawasaki/Mitsubishi
Class:	Yushio
Number:	SS-582
Mission:	attack submarine
Length:	76m (251ft)
Beam:	9.9m (33ft)
Displacement:	2500 tonnes (2540 tons)
Speed:	20 knots
Operating Depth:	300m (990ft)
Maximum Depth:	400m (1320ft)
Crew:	75
Nuclear Weapons:	none
Conventional Weapons:	533mm (21in) torpedo, Harpoon
Sonar:	ZQQ-4 sonar suite
Navigation:	ZPS-6 radar
Powerplant:	diesel-electric, 7200 shp
Date Commissioned:	1989

HNLMS BRUINVIS

The Walrus 2 class attack submarine is the most modern submarine in the Royal Netherlands Navy (RNLN). The class has been in service since the early 1980s, with the overhauled HNLMS *Bruinvis* representing the last to be launched in 1995. It was brought into service to replace the ageing Zwaardvis class attack vessel. It shares some similarities with its predecessor, having as it does the same general arrangement and hull form as the Zwaardvis 2 class. It also has the same system layout, but differs in having improved diving depth, reduced crew numbers and increased reliability, availability and safety. The operational range of the RNLN submarines lies mainly in the Eastern Atlantic Ocean, the North Sea and the Norwegian Sea, but also in the Mediterranean. Their missions are directed at anti-surface and anti-submarine warfare, carrying out surveillance, special operations and the laying of mines. This modern and sophisticated class contributes an important part to missions nationally and internationally agreed upon, as is the norm for the Dutch armed forces in general. The Dutch Navy takes an active role in participating in NATO naval exercises, and HNLMS *Bruinvis* has taken part in many. She has also recently returned from active duty as part of the US led anti-terrorist operation in Afghanistan, Enduring Freedom.

SPECIFICATIONS

Builder:	Droogdok Maatschappij B.V
Class:	Walrus 2
Number:	S-810
Mission:	attack submarine
Length:	67.7m (223ft)
Beam:	8.4m (28ft)
Displacement:	2800 tonnes (2844 tons)
Speed:	21 knots
Operating Depth:	300m (990ft)
Maximum Depth:	400m (1320ft)
Crew:	49
Nuclear Weapons:	none
Conventional Weapons:	533mm (21in) torpedo, mines
Sonar:	Thomson-Sintra TSM 2272
Navigation:	Signaal/Decca 1229
Powerplant:	diesel-electric, 6200 shp
Date Commissioned:	1994

MORAY

The Moray class submarine is an effort by the Dutch Ministry of Defence to export a number of vessels based on the lessons learnt developing the Walrus and Zwaardvis class submarines. The term Moray stands for Multi Operational Requirement Affected Yield, and the design is essentially a concept framework from which individual buyers can choose the types of systems and configuration of the vessel. The basic hull can be ordered in five different sizes, which although generally retaining the same propulsion and performance, are tailored to varied roles to suit the buyer. The Moray class has six 533mm (21in) torpedo tubes that can launch a variety of weapons including Mk48 torpedoes and the Sub Harpoon surface-to-surface missile, shown above. Despite the potential with the design, however, the Moray has not had many countries interested in the design. The Egyptian Navy was keen on the design and signed a declaration of intent to purchase two boats, but since then progress has stalled. Similarly, Taiwan has expressed an interest in purchasing a number of diesel-electric attack submarines, but due to political difficulties and the Dutch policy of not selling contentious weapons to Taiwan against Chinese wishes this has curbed potential sales. It is a difficult time for submarine manufacturers, and it is uncertain whether the Moray class will ever be bought.

SPECIFICATIONS

Builder:	Droogdok Maatschappij B.V
Class:	Moray 1400 H
Number:	N/A
Mission:	multi-purpose submarine
Length:	64m (211ft)
Beam:	6.4m (21ft)
Displacement:	1800 tonnes (1829 tons)
Speed:	20 knots
Operating Depth:	300m (990ft)
Maximum Depth:	360m (1188ft)
Crew:	32
Nuclear Weapons:	none
Conventional Weapons:	533mm (21in) torpedo, Harpoon
Sonar:	SIASS sonar suite
Navigation:	unknown
Powerplant:	diesel-electric AIP
Date Commissioned:	N/A

ROMEO

The North Korean Navy is the most secretive navy in the world. Determining the state of its fleet is exceptionally difficult and any information is sketchy at best. Even to Western intelligence agencies, names, numbers and bases are not known, and detailed quantities in service are uncertain. This makes determining the Democratic People's Republic of Korea's submarine capabilities extremely difficult. The Romeo class boats operated by the North Korean Navy are indigenously built at either the Nampo or Wonsan shipyards to a 1950s Soviet design. The Romeo class has long been considered obsolete by the Russians, and by modern standards they are archaic in design and capability. Nonetheless, the Romeo class submarines, despite being outdated and slow, are sufficiently capable of blocking sea lanes, but probably only for a limited period. These vessels could also attack South Korea surface vessels, emplace mines anywhere within South Korean maritime territory, or secretly infiltrate commandos into the South. Though North Korea's submarine capability is more readily associated with covert midget submarines, the Romeo class vessels form the backbone of the fleet. There are no plans to replace these aged boats with more modern equivalents, though this is probably down to a lack of funds and the political situation rather than a lack of will.

SPECIFICATIONS

Builder:	Nampo/Wonsan shipyards
Class:	Romeo
Number:	unknown
Mission:	coastal patrol submarine
Length:	76.6m (249ft)
Beam:	6.3m (21ft)
Displacement:	1700 tonnes (1727 tons)
Speed:	13 knots
Operating Depth:	170m (561ft)
Maximum Depth:	unknown
Crew:	50
Nuclear Weapons:	none
Conventional Weapons:	533mm (21in) torpedo, mines
Sonar:	Tamir-5L active
Navigation:	unknown
Powerplant:	diesel-electric, 2700 shp
Date Commissioned:	unknown

SANG-O

The Sang-O class midget submarines of the Democratic People's Republic of Korea Navy are supposedly designed along the lines of the famous Yugoslavian midget submarines. The primary offensive mission of the navy is supporting army actions against South Korea, particularly by inserting small-scale amphibious operations, including special operations forces, along the coast. The Sang-O class boats have also been used in recent years to harass South Korean shipping, though the frequency with which this happens has diminished slightly since the late 1990s. The Sang-O class coastal submarines belong to the Special Naval Infiltration Unit of the Reconnaissance Bureau of the General Staff Department, the Korean People's Army. This Bureau and the Korean Workers' Party's Liaison Office are primarily responsible for intelligence and other covert actions in South Korea. An infiltration team typically numbers three to five men, consisting of two escorts and one to three agents. On insertion missions the escorts are responsible for securing a landing site and then delivering the agents safely ashore. During exfiltration, escorts meet the agents on shore and escort them back to the submarine. How frequently Sang-O boats perform this function is unclear, though a few have been intercepted by Japanese and South Korean forces in recent times.

SPECIFICATIONS

Builder:	Nampo/Wonsan
Class:	Sang-O
Number:	unknown
Mission:	covert infiltration submarine
Length:	35.5m (116ft)
Beam:	3.8m (12ft)
Displacement:	277 tonnes (281 tons)
Speed:	9 knots
Operating Depth:	100m (330ft)
Maximum Depth:	200m (660ft)
Crew:	19 + 6 special operations forces
Nuclear Weapons:	none
Conventional Weapons:	533mm (21in) torpedo, mines
Sonar:	unknown
Navigation:	unknown
Powerplant:	diesel-electric, 800 shp
Date Commissioned:	unknown

YUGO

The Yugo class midget submarine is synonymous with the naval activities of North Korea, yet little is actually known about them. Yugoslavia became synonymous with the design and construction of miniature submarines, and is known to have exported the idea around the world over the decades. In the case of the North Korea Yugo, however, the name Yugo, implying Yugoslav design origin, appears to be a deception, as the vessels appear to bear no resemblance to any of the relatively sophisticated Yugoslav midget submarine designs. Such discoveries have been made in part because of the sporadic capture of these vessels by South Korean warships. North Korea lost another Yugo class saboteur infiltration submarine to operational ineptitude in 1998. According to South Korean sources the craft was found to be of primitive design and without the expected torpedo armament, though it is known that some Yugo submarines are armed with torpedoes. Though the North Korean Navy has a large number of these vessels, some estimates totalling over 50 examples, they are of such a primitive design that in wartime they would stand little chance if discovered by South Korean warships. However, finding such a small craft is difficult even with the most modern sensors, thus one could expect that with a measure of luck, some would get through to deliver their agents.

SPECIFICATIONS

Builder:	unknown
Class:	Yugo
Number:	unknown
Mission:	covert infiltration submarine
Length:	20m (66ft)
Beam:	2m (7ft)
Displacement:	25 tonnes (25.4 tons)
Speed:	4 knots
Operating Depth:	unknown
Maximum Depth:	unknown
Crew:	2 + 7 special operations forces
Nuclear Weapons:	none
Conventional Weapons:	400mm (18in) torpedo
Sonar:	unknown
Navigation:	unknown
Powerplant:	diesel-electric, 160 shp
Date Commissioned:	1960

KNM ULA

The Norwegian Submarine Force consists of six improved Kobben class and six Ula class submarines. Both these classes of vessel are designed for littoral operations and are essentially patrol boats. The Royal Norwegian Navy takes a full and active part in NATO training exercises, and the Ula class recently took part in an extensive exercise in the North Sea called Sorbet Royal 2002. In the wake of the *Kursk* disaster, modern naval operators have begun to explore more rigorously their response doctrines to submarines in distress. During the exercise, the *Utsira*, *Ula*'s sister ship, was deliberately taken to the ocean floor and 18 volunteers of her crew confined to the smallest of the boat's two compartments. Mock rescue operations were then undertaken to rescue these "trapped" crew members. The exercise was a useful experiment in testing NATO's disaster-response doctrine, and the lessons learned will save the lives of any sailors unfortunate enough to be trapped in a sunken submarine. In wartime, the Ula class can gather information on the activities of the enemy, covertly lay mines or can actively destroy the opposition's naval assets. However, in reality the Norwegian Navy will never have to fight without being part of some form of international force, and thus much of its training is geared towards operating effectively with its NATO and UN allies.

SPECIFICATIONS

Builder:	Thyssen Nordseewerke
Class:	Ula
Number:	S-300
Mission:	patrol submarine
Length:	59m (195ft)
Beam:	5.4m (18ft)
Displacement:	1150 tonnes (1168 tons)
Speed:	23 knots
Operating Depth:	unknown
Maximum Depth:	unknown
Crew:	21
Nuclear Weapons:	none
Conventional Weapons:	533mm (21in) torpedo, mines
Sonar:	Atlas Electronik CSU83
Navigation:	Kelvin Hughes Type 1007
Powerplant:	diesel-electric, 1900 shp
Date Commissioned:	1992

PNS HURMAT

The Pakistani Navy acquired two French-built Agosta class attack submarines like the Malaysian Navy one shown above in the late 1970s. They became known as the Hashmat class in Pakistani service. They were originally intended for the South African Navy but were sold to Pakistan before delivery. The procurement of these vessels was a marked leap forward for Pakistan's naval capabilities, since the Agosta submarines were state-of-the-art boats in their day. They have been superceded by the newer Agosta 90Bs, but will remain in Pakistani service for some years to come. Designed and built in the early 1970s, the Agosta class is a versatile and capable diesel-electric submarine. The design was extensively refitted in the late 1980s to fire the Exocet missile. This transformed the Agosta into a ship-killer as well as an anti-submarine vessel, making it sought after by many developing nations. It is a relatively cheap and effective solution, but is now decidedly outdated compared to modern diesel-electric attack submarines. The Pakistani variant of the Agosta is on constant alert, as tensions between Pakistan and its larger neighbour India are almost constantly fraught. Pakistan is continuing in its attempts to procure more modern vessels, obtaining as it has recently the Agosta 90B. However, the Hashmat class is expected to remain in service, but primarily in the patrol role.

SPECIFICATIONS

Builder:	DCN International
Class:	Hashmat (Agosta)
Number:	S-136
Mission:	attack/patrol submarine
Length:	67.6m (223ft)
Beam:	6.8m (22ft)
Displacement:	1725 tonnes (1752 tons)
Speed:	20 knots
Operating Depth:	250m (825ft)
Maximum Depth:	300m (990ft)
Crew:	54
Nuclear Weapons:	none
Conventional Weapons:	533mm (21in) torpedo, Exocet
Sonar:	Thomson DUUA 2 active/passive
Navigation:	Thomson CSF DRUA 33
Powerplant:	diesel-electric, 4600 shp
Date Commissioned:	1980

PNS KHALID

The Agosta 90B class submarine is designed by the French company DCN, and is currently in service with the French, Spanish and Pakistani navies. The Agosta 90B is an improved version of the original Agosta, featuring higher performance and a new combat system. The new submarine features a higher level of automation, which means that the ship's crew has been reduced from 54 to 36 sailors. Other improvements include a new battery for increased range, a deeper diving capability of 350m (1155ft), made possible by the use of HLES 80 steel, and a reduced acoustic signature through the installation of new suspension and isolation systems. Three Agosta 90Bs were ordered by the Pakistani Navy in September 1994. The first, PNS *Khalid*, was built at DCN's Cherbourg yard and was commissioned in 1999. The second, PNS *Saad*, was assembled at Karachi Naval Dockyard and launched in August 2002. The third, PNS *Hamza*, which is being constructed and assembled indigenously at Karachi, is scheduled to be completed by 2005. Work on the vessel was interrupted following a terrorist attack in May 2002, which killed 11 French engineers, but has since restarted. PNS *Hamza* is being fitted with the MESMA Air Independent Propulsion system (AIP). The MESMA AIP system will eventually be retro-fitted to PNS *Khalid* and PNS *Saar*.

SPECIFICATIONS

Builder:	DCN International
Class:	Khalid (Agosta 90B)
Number:	S-137
Mission:	attack submarine
Length:	67.6m (223ft)
Beam:	6.8m (22ft)
Displacement:	1760 tonnes (1788 tons)
Speed:	17 knots
Operating Depth:	300m (990ft)
Maximum Depth:	350m (1155ft)
Crew:	36
Nuclear Weapons:	none
Conventional Weapons:	533mm (21in) torpedo, Exocet
Sonar:	TSM 223 sonar suite
Navigation:	Thales I-band radar
Powerplant:	diesel-electric, 3600 shp
Date Commissioned:	1999

ARKHANGELSK

The Typhoon class boat TK-17 was scheduled to be scrapped, but has recently undergone a year-long refit and is still in active service, expecting to remain so until 2010. The crew petitioned the Navy Headquarters to adopt a new name in July 2002, and it is thus now known as *Arkhangelsk*. This is not an unusual practice in Russian naval circles, as ships can be renamed at any time. This is in stark contrast to most navies around the world. It is a Typhoon class submarine, identical to all the others with the exception that the *Arkhangelsk* is still in operational service. TK-17 was involved in an accident in the Pacific in 1992, where a missile exploded during testing. The *Arkhangelsk* was extensively damaged in the explosion, but has since been repaired. The Typhoon class submarine is of a multi-hulled design. Five inner hulls are situated inside a superstructure of the two parallel main hulls. The superstructure is coated with sound-absorbent tiles. These drastically reduce another submarine's ability to detect the Typhoon. There are 19 compartments, including a strengthened module which houses the main control room and electronic equipment compartment. The design includes features to enable it to both travel under ice and for ice-breaking. Indeed, a great many of the "cat and mouse" games of the Cold War took place under the polar ice-caps.

SPECIFICATIONS

Builder:	Sevmash
Class:	Typhoon
Number:	TK-17
Mission:	ballistic missile submarine
Length:	172m (564ft)
Beam:	23m (76ft)
Displacement:	2264 tonnes (2300 tons)
Speed:	25 knots
Operating Depth:	350m (1155ft)
Maximum Depth:	500m (1650ft)
Crew:	150
Nuclear Weapons:	20 x SS-N-24 SLBM
Conventional Weapons:	630mm (28in) torpedo, ASROC
Sonar:	Shark Gill sonar suite
Navigation:	unknown
Powerplant:	nuclear reactors, 100,000 shp
Date Commissioned:	1987

AS-19

The Uniform class submarine is a small, deep-diving nuclear submarine. There is not much open source information about this vessel, which suggests that it is used in some form of covert operations, though the manner of which one can only speculate. It is most likely to be employed in "ocean engineering", which is a well-known intelligence euphemism for covert seafloor operations. It is believed to be capable of diving up to depths of well over 910m (3000ft), though obviously precise figures are not available. To achieve this depth it is almost certain that the hull is made of some kind of titanium alloy, perhaps similar to the type used on Sierra class attack submarines. It is also said to be capable of travelling at over 30 knots, which is a remarkable speed for a craft of its size. In addition to its shaft-driven propulsion, the Uniform class is also fitted with side thrusters, which give it high manoeuvrability whilst partaking in its "research", and allows it to hover over one spot. It is unknown whether the Uniform class can disembark any kind of special operations forces, but it would not be too surprising to learn that it did, if that information ever became public. In essence this vessel is probably similar in role to the United States Navy NR-1 deep-diving research vessel, though the AS-19 is much larger, but the reason for its extra size is top secret.

SPECIFICATIONS

Builder:	United Admiralty Shipyard 196
Class:	Uniform
Number:	AS-19
Mission:	special missions submarine
Length:	69m (226ft)
Beam:	7m (23ft)
Displacement:	1580 tonnes (1605 tons)
Speed:	30 knots
Operating Depth:	unknown
Maximum Depth:	910m (3000ft)
Crew:	36
Nuclear Weapons:	none
Conventional Weapons:	none
Sonar:	high-frequency active arrays
Navigation:	unknown
Powerplant:	nuclear reactor, 6000 shp
Date Commissioned:	1995

AS-35

The Paltus class submarine is the Russian equivalent of the United States Navy NR-1 deep-diving research vessel. It is powered by a small nuclear reactor, but is capable of only six knots. In addition to the main propulsion, it is also fitted with side thrusters to increase its manoeuvrability when on the seafloor. Like the Uniform class deep-diving submarine, the nomenclature of "research vessel" is a little erroneous, because without doubt these vessels are used for covert operations of one form or another. These could include the emplacement of temporary or fixed sensor arrays on the ocean floor and their subsequent maintenance, the investigation of unknown objects, the provision of deep-sea targeting for other vessels or even involvment in rescue operations with stricken submarines. It is also possible that this vessel could deliver special operations forces into highly defended areas. It is believed that the Yankee Stretch conversion of a Yankee class SSBN serves as a mothership for transport and support of these craft, taking them across the vast distances of the oceans before deploying them close to their target areas. This allows the Paltus to be deployed theoretically anywhere in the world at relatively short notice. Numbers of exactly how many Paltus class submarines exist are not readily available; indeed, there may well be only a single craft in service.

SPECIFICATIONS

Builder:	United Admiralty Shipyard 196
Class:	Paltus
Number:	AS-35
Mission:	special missions submarine
Length:	53m (174ft)
Beam:	3.8m (13ft)
Displacement:	730 tonnes (741 tons)
Speed:	6 knots
Operating Depth:	unknown
Maximum Depth:	unknown
Crew:	14
Nuclear Weapons:	none
Conventional Weapons:	none
Sonar:	unknown
Navigation:	1000m (3280ft)
Powerplant:	nuclear reactor, 300 shp
Date Commissioned:	1995

BELGORAD

The Akula II class attack submarine is the latest member of the Akula family to begin entering into service with the Russian Navy. This vessel was the response from the Russian authorities to the realization that US submarines, including the Los Angeles class and the next generation of attack submarines, had the edge over any existing Russian vessel. However, production of the new design has been undertaken at a snail's pace, and there is speculation that the majority of the improved vessels may never see service. The design of the Akula II differs slightly from its predecessors in the fact that it is 4m (13ft) longer. The reason for this is widely believed to be the incorporation of a quieter propulsion system and additional sound-dampening equipment. Furthermore, the Akula II has shifted from being a simple strategic attack submarine to a more general-purpose one. This shift in the doctrine of how the vessel is used means that the breadth of missions that the Akula II can take part in has increased. This is largely driven by the necessity to meet a wide range of potential challenges in an effective yet cost-effective manner. The Akula II lacks none of the required attributes to continue being a formidable submarine, yet the funds needed to alter the operational doctrine and retrain the crews for operations other than anti-submarine actions are decidedly lacking.

SPECIFICATIONS

Builder:	Sevmash
Class:	Akula II
Number:	unknown
Mission:	attack submarine
Length:	112m (369ft)
Beam:	13.5m (45ft)
Displacement:	9500 tonnes (9652 tons)
Speed:	32 knots
Operating Depth:	475m (1570ft)
Maximum Depth:	545m (1800ft)
Crew:	51–62
Nuclear Weapons:	none
Conventional Weapons:	10 x tubes (533mm, 650mm)
Sonar:	MGK Skat, mine detection sonar
Navigation:	Medvyedista-945
Powerplant:	nuclear reactor, 43,000 shp
Date Commissioned:	2001

CHELYABINSK

The Oscar II cruise-missile attack submarine is one of the most advanced of its kind in the world. It is also one of the largest submarines in service, displacing almost 24,000 tonnes (24,384 tons) submerged. During the Cold War, the Soviet Union was deeply afraid of the US aircraft carrier battle groups, and the Oscar class attack submarine was designed to counter this threat. Although the Russian Navy has been forced to scrap large numbers of its submarines, the Oscar II submarines are so important to the Russian Navy that they have received sufficient funds to be maintained in effective operating order. This said, it is believed that three or four vessels are waiting to be scrapped. The Oscar II is more capable than its Oscar I predecessor, and is 10m (33ft) longer, possibly to incorporate a quieter propulsion system. The *Chelyabinsk* is capable of travelling over 30 knots whilst submerged, and carries a compliment of 24 SS-N-19 anti-shipping missiles with a range of 550km (343 miles). The missiles, which are launched while the submarine is submerged in order to achieve surprise and remain undetected, are fired from tubes fixed at an angle of approximately 40 degrees. It can also fire torpedoes and shorter-range anti-shipping missiles from its four 533mm (21in) bow torpedo tubes. The Oscar II class are still awesome submarines.

SPECIFICATIONS

Builder:	Sevmash
Class:	Oscar II
Number:	K-442
Mission:	cruise-missile submarine
Length:	154m (508ft)
Beam:	18.2m (60ft)
Displacement:	24,000 tonnes (24,384 tons)
Speed:	32 knots
Operating Depth:	300m (990ft)
Maximum Depth:	600m (1980ft)
Crew:	94–118
Nuclear Weapons:	none
Conventional Weapons:	SS-N-19, 533mm (21in) torpedo
Sonar:	Shark Gill sonar
Navigation:	Snoop Pair radar
Powerplant:	nuclear reactor, 90,000 shp
Date Commissioned:	1991

DANIIL MOSKOVSKIY

An improved version of the Victor II, the Victor III was an interim Soviet effort to apply some level of silencing to their submarines. The hull was lengthened by nearly 6m (20ft) to accommodate the rafting and sound insulation for the turbine machinery. The design also features improvements in electronics, navigation systems, and radio and satellite communication systems, accommodated in the additional hull space forward of the sail. All Victor class boats are double-hulled, as is the case in almost all Russian submarines. The outer hull is coated with anti-hydro-acoustic materials to reduce the possibility of detection. The outer hull of the Victor III is made partly from light alloys, and is distinguishable by a high stern fin fitted with a towed array dispenser, the first Soviet submarine to be fitted with a towed array. A total of 26 units were constructed during the Cold War. The Victor class submarines were designed to engage enemy ballistic missile submarines, anti-submarine task forces, and to protect friendly vessels and convoys from enemy attacks. A contemporary of the American Sturgeon class, they were significantly faster but also had much higher noise levels; indeed, the first two designs made no significant effort to reduce noise emissions. Despite its relative age and lack of sophistication, the Victor III is still in service.

SPECIFICATIONS

Builder:	Admiralty Shipyard
Class:	Victor III
Number:	K-388
Mission:	attack submarine
Length:	107m (353ft)
Beam:	11m (34ft)
Displacement:	7250 tonnes (7366 tons)
Speed:	29 knots
Operating Depth:	300m (990ft)
Maximum Depth:	400m (1320ft)
Crew:	85–100
Nuclear Weapons:	none
Conventional Weapons:	533mm (21in) torpedo, mines
Sonar:	Rubikon sonars
Navigation:	Medvyedista-671
Powerplant:	nuclear reactor, 30,000 shp
Date Commissioned:	1984

DELFIN

The original Kilo Class submarine was designed for anti-submarine and anti-ship warfare in the protection of naval bases, coastal installations and sea lanes, and also for general reconnaissance and patrol missions during the Cold War. Despite its relative age, the Kilo is considered to be one of the quietest diesel submarines in the world. The Project 636 design, which is known in NATO as Improved Kilo, is a generally enhanced development of the original design. The Improved Kilo is actively promoted for the world market by the Rosvoorouzhenie state-owned company, and has been successfully exported to a number of countries including Iran and India. The main differences between the two variations are that the Improved Kilo has better range, firepower, acoustic characteristics and reliability. The hull has also been lengthened by around 1.2m (54in). The Improved Kilo is equipped with six 533mm (21in) forward torpedo tubes situated in the nose of the submarine, and carries 18 torpedoes with 6 in the torpedo tubes and 12 stored on the racks. The submarine can also carry 24 mines with 2 in each of the 6 tubes and 12 on the racks. The Improved Kilo also incorporates a computer-controlled torpedo system with a quick-loading device. It takes only 15 seconds to prepare standby torpedo tubes for firing.

SPECIFICATIONS

Builder:	Admiralty Shipyard
Class:	Improved Kilo
Number:	B-880
Mission:	ASW submarine
Length:	73.8m (242ft)
Beam:	9.9m (32ft)
Displacement:	3126 tonnes (3176 tons)
Speed:	20 knots
Operating Depth:	250m (825ft)
Maximum Depth:	300m (990ft)
Crew:	52
Nuclear Weapons:	none
Conventional Weapons:	533mm (21in) torpedo, mines
Sonar:	Rubikon active/passive sonar
Navigation:	GPS navigation system
Powerplant:	diesel-electric, 5500 shp
Date Commissioned:	1993

KARELIA

The development of the 667BDRM project, more commonly known as the Delta IV class SSBN, ensured that several measures were included to improve its overall capability over its Delta III predecessor. The most significant of these was to reduce its noise level. The gears and equipment were moved and placed on a common base isolated from the pressure hull, and the power compartments were also isolated. The efficiency of the anti-hydro-acoustic coatings of the light outer hull and inner pressure hulls was further improved. A newly designed five-bladed propeller with improved hydro-acoustic characteristics was also introduced. Delta IV submarines are equipped with the TRV-671 RTM missile-torpedo system that has four 533mm (21in) torpedo tubes. Unlike the Delta III, it is capable of using all types of torpedoes, anti-submarine torpedo-missiles and anti-hydro-acoustic devices. The battle-management system, Omnibus-BDRM, controls all combat activities, processes data and commands the torpedo weapons. The Shlyuz navigation system provides for the improved accuracy of the missiles and is capable of stellar navigation at periscope depths. The navigation system also employs two floating antenna buoys to receive radio messages, target destination data and satellite navigation signals at great depth.

SPECIFICATIONS

Builder:	Sevmash
Class:	Delta IV 'Delfin'
Number:	K-18
Mission:	ballistic missile submarine
Length:	167m (551ft)
Beam:	12m (39ft)
Displacement:	18,200 tonnes (18,941 tons)
Speed:	24 knots
Operating Depth:	320m (1056ft)
Maximum Depth:	400m (1320ft)
Crew:	130
Nuclear Weapons:	6 x R-29 RM SLBM
Conventional Weapons:	533mm (21in) torpedo
Sonar:	Skat VDRM
Navigation:	Shlyuz system
Powerplant:	nuclear reactor, 40,000 shp
Date Commissioned:	1991

KASHALOT

The Akula class attack submarine is one of the Russian Navy's most successful and effective weapons systems. Designed and built during the latter years of the Cold War, the Akula was the replacement for the Sierra class submarine. It is made from low-magnetic steel, which was the cheaper and more easily produced alternative to the titanium hull sported by its Sierra class predecessor. This compromised on the diving depth, but was a more practical choice. Like almost all Russian submarines, it is designed with a double hull, in this case with significant distance between the two hulls, giving the inner hull a greater degree of protection in the event of attack. It also has a distinctive aft fin that makes it easily recognizable. Its primary role is to attack surface shipping using either its 533mm (21in) or 650mm (28in) torpedoes, or its SS-N-15 Starfish and SS-N-16 Stallion anti-shipping missiles. It can further attack coastal positions using its Granat cruise missiles which are fired from the 533mm (21in) torpedo tubes. The Akula class can also defend itself from the air with a portable shoulder-launched Strela SA-N-5 surface-to-air missile. This can be fired by a crew member from the deck of the submarine or from the top of the conning tower. The Akula class submarine is a fearsomely armed and powerful vessel, and is still one of the most effective vessels of its kind in the world.

SPECIFICATIONS

Builder:	Sevmash
Class:	Akula
Number:	K-322
Mission:	attack submarine
Length:	108m (356ft)
Beam:	13.5m (45ft)
Displacement:	9100 tonnes (9246 tons)
Speed:	32 knots
Operating Depth:	400m (1320ft)
Maximum Depth:	550m (1800ft)
Crew:	51–62
Nuclear Weapons:	none
Conventional Weapons:	533mm (21in) torpedo, Granat
Sonar:	503-M Skat sonar suite
Navigation:	Medvyedista-945
Powerplant:	nuclear reactor, 43,000 shp
Date Commissioned:	1987

KRAB

The original Sierra I class attack submarine was introduced into Russian service at the height of the Cold War in the mid-1980s, and was seen as a match for the US Los Angeles class submarine. It is generally comparable in performance to early Los Angeles class vessels, though with an arguably superior non-acoustic detection system and integrated acoustic countermeasures system. However, the Improved Los Angeles class boats and subsequent modern designs outstripped the Sierra I's capabilities, and production was switched to the more modern Sierra II class boats. Thus there are only a couple of Sierra Is in service, if indeed they are in operational order. The Sierra design incorporates Cluster Guard anaechoic tiles on the outer hull, which serves to reduce noise levels. Cluster Guard is the official NATO reporting name for sound-absorbing material on newer Soviet submarines. The material is sometimes referred to as tiles because of its block-like configuration, and it is apparently able to reduce the effectiveness of active anti-submarine sonars and active-acoustic torpedo guidance. This technology is featured on almost all newly built submarines, and is seen as an effective countermeasure to the advance of sonar detection systems. The Sierra I was behind the times before it was launched, and is outdated in the twenty-first century.

SPECIFICATIONS

Builder:	Sevmash
Class:	Sierra I
Number:	K-276
Mission:	attack submarine
Length:	107m (353ft)
Beam:	11m (33ft)
Displacement:	10,100 tonnes (10,261 tons)
Speed:	35 knots
Operating Depth:	696m (2300ft)
Maximum Depth:	803m (2650ft)
Crew:	61
Nuclear Weapons:	nuclear depth charges
Conventional Weapons:	533mm (21in) torpedo, mines
Sonar:	Skat sonar suite
Navigation:	Medvyedista-945
Powerplant:	nuclear reactor, 50,000 shp
Date Commissioned:	1987

K-403 YANKEE POD

The K-403 Yankee Pod class submarine is a Yankee class SSBN that has been converted from its ballistic missile role to something more useful for Russia's naval forces. It was subject to the same standard conversion as most of the Yankee class SSBNs, in that it had its missile compartment cut out as part of international arms control agreements. The original Yankee class SSBNs were an integral part of the Soviet Union's nuclear forces, and 34 individual boats served the Russian Navy for over 30 years. They were armed with the R-27 ballistic missile which was able to strike at targets up to 4800km (3000 miles) away. Like the Yankee Stretch, but unlike most Yankee class boats, it was fully converted to another role. This is believed to be as a nuclear-powered trials submarine. It is unclear what experimental weaponry or systems the Yankee Pod has been involved in testing since being commissioned in 1980, which is unsurprising given the nature of the intense secrecy that surrounds submarine operations. However, it is believed that this vessel is currently testing the Irtysh-Amfora sonar suite intended for the new Sevmash class attack submarine. Since a trials submarine is not often subject to the same stresses and strains of fully operational vessels, the Yankee Pod is expected to continue serving as a trials submarine for some time to come.

SPECIFICATIONS

Builder:	Sevmash
Class:	Yankee Pod
Number:	K-403
Mission:	trials submarine
Length:	134m (440ft)
Beam:	12m (39ft)
Displacement:	10,100 tonnes (10,261 tons)
Speed:	27 knots
Operating Depth:	300m (990ft)
Maximum Depth:	400m (1320ft)
Crew:	110
Nuclear Weapons:	none
Conventional Weapons:	none
Sonar:	Irtysh-Amfora sonar suite
Navigation:	Tobol navigation system
Powerplant:	nuclear reactor, 52,000 shp
Date Commissioned:	1980

K-411 YANKEE STRETCH

The K-411 Yankee Stretch submarine is a conversion from a Yankee class SSBN that had been in service with the Russian Navy since 1970. However, as with almost all of the Yankee class boats, K-411 was removed from operational status and her missile compartments cut out to comply with arms control agreement ceilings. Whereas the majority of the Yankee class were decommissioned, a small number were converted. The Yankee Stretch was converted by replacing the missile compartment with an extended hull section. This is widely agreed to be because the K-411 has been extensively modified to serve as the mothership to a very small special operations submarine, the AS-35 Paltus class. The principle is the same as the dry-dock shelter system used by some US submarines, such as the USS *Kamehameha*, in that a larger carrier vessel provides covert support for subaqua special forces operations. This is quite a shift in operational doctrine for the Soviet armed forces, but is a reflection of the efforts being made in the services to meet the new challenges of warfare in the twenty-first century. It is understood that this vessel carries no armament, and is thus reliant on the use of stealth for protection, or it is protected by other armed submarines. It is believed that K-411 is part of the Russian Northern Fleet, and is still operational.

SPECIFICATIONS

Builder:	Sevmash
Class:	Yankee Stretch
Number:	K-411
Mission:	special missions submarines
Length:	160m (525ft)
Beam:	12m (39ft)
Displacement:	11,600 tonnes (11,785 tons)
Speed:	28 knots
Operating Depth:	300m (990ft)
Maximum Depth:	400m (1320ft)
Crew:	110
Nuclear Weapons:	none
Conventional Weapons:	none
Sonar:	unknown
Navigation:	Tobol navigation system
Powerplant:	nuclear reactors, 52,000 shp
Date Commissioned:	1990

KURSK

The submarine at the heart of one of Russia's most tragic maritime accidents was the Oscar II class cruise missile submarine, K-141 *Kursk*. She sank with all hands after an explosion in her torpedo compartment. Despite early optimism that many of the crew had survived following the explosion, it has since been established that the majority of the crew either died instantly or within hours of the accident. A small number of crewmen did survive the initial explosion but were trapped in the aft of the boat and died shortly from carbondioxide poisoning. Nonetheless, Norwegian and British rescue vessels raced to the scene, but due to inhospitable weather and sea conditions they were unable to achieve any success. On 21 August 2000, Chief of Staff of the Russian Northern Fleet Mikhail Motsak pronounced the *Kursk* flooded and its whole crew dead. As is often the case with such events, claims, counter-claims and conspiracy theories are in abundance. Some Russian sources claim that the *Kursk* was involved in a collision with a US or Royal Navy vessel, or even that the *Kursk* hit a World War II mine, yet an explosion in the torpedo room is now the accepted reason for the *Kursk*'s sinking. The loss of the *Kursk* cast a dark shadow over the entire Russian Navy, and indeed over most of Russian society, and it was seen as the direct result of defence budget cuts.

SPECIFICATIONS

Builder:	Sevmash
Class:	Oscar II
Number:	K-141
Mission:	cruise-missile submarine
Length:	154m (508ft)
Beam:	18.2m (60ft)
Displacement:	24,000 tonnes (24,384 tons)
Speed:	32 knots
Operating Depth:	300m (990ft)
Maximum Depth:	600m (1980ft)
Crew:	94–118
Nuclear Weapons:	none
Conventional Weapons:	533mm (21in) torpedo, Granat
Sonar:	Shark Gill sonar
Navigation:	Snoop Pair radar
Powerplant:	nuclear reactor, 90,000 shp
Date Commissioned:	1994

NOVOMOSKOVSK

The 667BDRM Delta IV class submarine was constructed parallel to the Typhoon class, and is a follow-on design from its predecessor Delta III class. The Delta IV class is one of Russia's most modern and important submarines. In comparison with the Delta III submarine the diameter of the pressure hull was increased and the bow was lengthened. As a result the displacement of the submarine was increased by 1200 tonnes (1219 tons) and it was lengthened by 12m (39ft). The Delta IV class vessels employ the D-9RM ballistic missile launch system and carry 16 R-29RM liquid-fuelled, submarine-launched ballistic missiles (SLBMs). Each missile carries four multiple independently targetable re-entry vehicles (MIRVs). Each missile is capable of hitting four different targets over 8000km (5000 miles) away, and each of the four MIRV warheads has the explosive power equivalent to over one million tonnes (1.016 million tons) of TNT. There is enough explosive power on a single Delta IV class SSBN to destroy every single state capital in the United States in a matter of minutes. Unlike previous modifications, the Delta IV submarine is able to fire missiles in any direction from a constant course in a circular sector. The underwater firing of the ballistic missiles can be conducted at a depth of 55m (181ft) while cruising at a speed of 6–7 knots.

SPECIFICATIONS

Builder:	Sevmash
Class:	Delta IV
Number:	K-407
Mission:	ballistic missile submarine
Length:	167m (551ft)
Beam:	12m (39ft)
Displacement:	18,200 tonnes (18,941 tons)
Speed:	24 knots
Operating Depth:	320m (1056ft)
Maximum Depth:	400m (1320ft)
Crew:	130
Nuclear Weapons:	16 x R-29 RM SLBM
Conventional Weapons:	533mm (21in) torpedo, mines
Sonar:	Skat VDRM
Navigation:	Shlyuz system
Powerplant:	nuclear reactor, 40,000 shp
Date Commissioned:	1992

PANTERA

The Akula class attack submarine was one of the most feared Russian vessels during the last years of the Cold War. It had been assumed that the Russians could not match the Western powers in terms of technological capability. Whereas this assumption was generally true across most military technical areas, the introduction of the Akula clearly showed that this was not the case for submarines. It came as something of a shock to US intelligence that the Soviet Union had been able to match US technology in its nuclear attack submarines, and the development of the Seawolf class SSN was a direct consequence of the arrival of the Akula class. The Akula is equipped with a sophisticated array of sensors, including the MGK-540 sonar system which provides automatic target detection in broad and narrow band modes by active sonar. It gives the range, relative bearing and range rate. Aside from the less stealthy active sonar, the system can also be used in a passive, listening mode for detection of hostile sonars. The system incorporates a powerful computer that can process and automatically classify targets as well as reject background acoustic noise sources and compensate for the variable acoustic conditions that are prevalent in the sea. The Akula is one of the few classes of boats that have not been left in such disrepair as to render them useless.

SPECIFICATIONS

Builder:	Sevmash
Class:	Akula
Number:	K-317
Mission:	attack submarine
Length:	108m (356ft)
Beam:	13.5m (45ft)
Displacement:	9100 tonnes (9245 tons)
Speed:	32 knots
Operating Depth:	475m (1570ft)
Maximum Depth:	545m (1800ft)
Crew:	51–62
Nuclear Weapons:	none
Conventional Weapons:	533mm (21in) torpedo, Strela
Sonar:	MGK-540 sonar suite
Navigation:	Medvyedista-945
Powerplant:	nuclear reactor, 43,000 shp
Date Commissioned:	1987

PSKOV

The Sierra II class attack submarine was one of Soviet Russia's greatest success stories, and represented the pinnacle of submarine design when it came into service. It was designed as a strategic attack submarine, capable of attacking surface vessels or coastal targets. It was a follow-on from its predecessor, the highly effective Sierra I, and had some major changes made to it in order to improve its capabilities. It differs from the Sierra I in that it has a better sonar system and has a reduced acoustic signature, making it less detectable. It is also around 5m (16.5ft) longer than the Sierra I and also has a larger blunt sail that is 6m (20ft) longer than the Sierra I sail. The increased hull size also provides improved living quarters for the crew and additional sound-dampening measures. It is also equipped with a new American-style spherical bow sonar, a definite break with tradition for the Russian armed forces. The inclusion of this bow sonar meant that there was no room for the torpedo tubes, which were thus moved farther towards the stern and angled out. The torpedo room was also modified to accommodate the Granat strategic cruise missile, giving the Sierra II far greater firepower and capability. As is often the case with specific details about Russian submarines, there is some debate over the number and state of Sierra II vessels still in service.

SPECIFICATIONS

Builder:	Sevmash
Class:	Sierra II
Number:	K-534
Mission:	attack submarine
Length:	113m (372ft)
Beam:	12m (40ft)
Displacement:	10,400 tonnes (10,566 tons)
Speed:	35 knots
Operating Depth:	697m (2300ft)
Maximum Depth:	803m (2650ft)
Crew:	61
Nuclear Weapons:	nuclear depth charges
Conventional Weapons:	533mm (21in) torpedo, Granat
Sonar:	Skat sonar suite
Navigation:	Medvyedista-945
Powerplant:	nuclear reactor, 50,000 shp
Date Commissioned:	1992

SEVERODVINSK

The Severodvinsk class attack submarine is the latest in the pipeline for the Russian Navy. It is a further derivative of the Akula class attack submarine that has been so successful, and is due to enter service sometime in the next few years. Work on the submarine began in 1992, but there has since then been an ongoing saga over the continued development. Work ceased in 1996 and there is no clear evidence that much has happened since, though estimates state that a budget version of the submarine has been considered as a compromise. US defence analysts have considered that the Severodvinsk class submarine will only be marginally quieter than the existing Akula attack submarine, thus one must question the necessity for another variation given the expense and the state of the Russian Navy. Furthermore, whilst Russian President Vladimir Putin professes a liking for naval matters in public, this is really a public-relations ploy and the fiscal reality is that the navy suffers perhaps more than any other branch of the beleaguered Russian armed forces in terms of underfunding, so there is a chance that the *Severodvinsk* may never be launched. If launched the *Severodvinsk* would be a very capable vessel. In addition to sophisticated sonar, it is designed with eight vertical launch tubes for cruise missiles, as well as four 563mm (25.6in) torpedo tubes.

SPECIFICATIONS

Builder:	Sevmash
Class:	Severodvinsk
Number:	none currently assigned
Mission:	attack submarine
Length:	111m (333ft)
Beam:	12m (40ft)
Displacement:	13,000 tonnes (13,208 tons)
Speed:	30 knots
Operating Depth:	455m (1500ft)
Maximum Depth:	606m (2000ft)
Crew:	50
Nuclear Weapons:	none
Conventional Weapons:	563mm (25.6in) torpedo, Granat
Sonar:	Irtysh-Amfora sonar suite
Navigation:	Myedvyeditsa-971
Powerplant:	nuclear reactor, 43,000 shp
Date Commissioned:	unknown

SEVERSTAL

The Typhoon ballistic missile nuclear-powered submarines are the largest submarines ever to be built, and were a defining feature of the Cold War. They were constructed at the Sevmash shipyard, on the White Sea near Archangel. The Typhoon class submarine is of multi-hulled design with five inner hulls situated inside a superstructure of the two parallel main hulls. The superstructure is coated with sound-absorbent tiles. There are 19 compartments, including a strengthened module which houses the main control room and electronic equipment compartment, which is above the main hulls behind the missile launch tubes. Maximum diving depth is officially 500m (1650ft), though it could be significantly more. The Typhoon class carries 20 of the RSM-52 intercontinental ballistic missiles designed by the Makayev Design Bureau. The NATO designation for the weapon is the SS-N-20 Sturgeon. Each missile consists of 10 independently targetable multiple re-entry vehicles (MIRVs), each with a 100 kiloton nuclear warhead. The missile has a range of 8300km (5100 miles) and is able to hit its target within 500m (1650ft). The Typhoon class has four 630mm (28in) torpedo tubes and two 533mm (21in) torpedo tubes, with a total of 22 anti-ship missiles and torpedoes of varying types. Though slated to be scrapped, the *Severstal* is still in service.

SPECIFICATIONS

Builder:	Sevmash
Class:	Typhoon
Number:	TK-20
Mission:	ballistic missile submarine
Length:	172m (564ft)
Beam:	23m (76ft)
Displacement:	33,800 tonnes (34,340 tons)
Speed:	25 knots
Operating Depth:	350m (1155ft)
Maximum Depth:	500m (1650ft)
Crew:	150
Nuclear Weapons:	20 x SS-N-24 SLBM
Conventional Weapons:	630mm (28in) torpedo, ASROC
Sonar:	Shark Gill sonar suite
Navigation:	unknown
Powerplant:	nuclear reactor, 100,000 shp
Date Commissioned:	1989

SYVATOY GIORGIY

The development of the 667BDR Delta III ballistic missile submarine began in 1972 at the Rubin Central Design Bureau for Marine Engineering. This strategic submarine is equipped with the D-9R launch system and 16 SS-N-18 missiles, and was the first submarine to be able to fire any number of missiles in a single salvo. The SS-N-18 missile was the first sea-based Soviet ballistic missile, each carrying 3–7 multiple independently targetable re-entry vehicles (MIRVs), with a range of up to 8000km (5000 miles) depending on the number of MIRVs deployed. The Delta III is equipped with the Almaz-BDR battle-management system and the Tobol-M2 inertial navigation system. The Delta III is also equipped with the Rubikon hydro-acoustic system. The Delta III SSBNs entered service in 1976, and by 1982 a total of 14 submarines were commissioned, all built at Sevmash shipyards. The operational lifetime of these submarines is estimated to be 20–25 years, so they are getting a little long in the tooth. In terms of capability the Delta III is behind its more modern counterparts, though the Russian Navy still retains a small number of these vessels, most of which are in reserve. There are no precise figures available in open-source intelligence as to how many Delta IIIs are still in operational order, though estimates put the number as low as four.

SPECIFICATIONS

Builder:	Sevmash
Class:	Delta III
Number:	K-433
Mission:	ballistic missile submarine
Length:	155m (511ft)
Beam:	11.7m (38ft)
Displacement:	10,600 tonnes (10,796 tons)
Speed:	24 knots
Operating Depth:	320m (1056ft)
Maximum Depth:	400m (1320ft)
Crew:	130
Nuclear Weapons:	16 x SS-N-18 SLBM
Conventional Weapons:	533mm (21in) torpedo, Vodopod
Sonar:	Rubikon sonar suite
Navigation:	Tobol M-2
Powerplant:	nuclear reactor, 60,000 shp
Date Commissioned:	1981

TIGR

The Akula class attack submarine has undergone a series of improvement refits and upgrades since its introduction. The first of these takes the Akula up to the Improved Akula I class. This vessel does not differ extensively from the Akula, but it does have a number of improvements. Along with the pre-existing six 630mm (28in) torpedo tubes, the upgrade includes the addition of two extra 533mm (21in) torpedo tubes to the existing compliment of four tubes, taking the number of available tubes to 10. This gives the Improved Akula I an awesome amount of firepower. There are only a handful of Improved Akula Is in the Russian Navy, with some estimates reckoning on only five or six boats split between the Northern and Pacific fleets. There are also rumoured to be improvements in the sound-dampening equipment in the Improved Akula, making it marginally quieter than the original. The vessel is at its quietest at a speed of 6–9 knots, but it is still not as quiet as the US Los Angeles class attack submarine at higher speeds. There is some debate amongst those who keep a close eye on Russian submarine developments about what the difference is between an Improved Akula and the Akula II. Ultimately it is difficult to tell since the majority of alterations on each version are in the heart of the vessel, obscured from view.

SPECIFICATIONS

Builder:	Sevmash
Class:	Improved Akula
Number:	K-157
Mission:	attack submarine
Length:	108m (356ft)
Beam:	13.5m (45ft)
Displacement:	9100 tonnes (9245 tons)
Speed:	32 knots
Operating Depth:	475m (1570ft)
Maximum Depth:	545m (1800ft)
Crew:	51–62
Nuclear Weapons:	none
Conventional Weapons:	630mm (28in) torpedo
Sonar:	Skat, MG-70 mine detection
Navigation:	Medvyedista-945
Powerplant:	nuclear reactor, 43,000 shp
Date Commissioned:	1994

VLADIKAVKAZ

The Kilo class submarine is a standard Soviet diesel-electric submarine design, which has also been produced extensively for export. It was designed with general coastal/littoral operations in mind, as opposed to deep-water operations. Despite the age of the design, and the improvements made to its Improved Kilo successor, the Kilo is still considered to be a very quiet vessel. The Russian Navy still has a number of Kilo class submarines in service, with estimates ranging around 14 on operational duty, and a further 7 in reserve, though the mix of standard and Improved Kilos is not known. The Russian fleet operates three variants of the Kilo, known as project 877: the basic 877, the 877K that has an improved fire-control system, and the 877M that can fire wire-guided torpedoes from two tubes. Though the Kilo class is no longer in production for the Russian Navy, it continues to be popular with other nations. The Kilo class has been a very successful export for the Russian Navy, and continues to generate a decent amount of much-needed income for the failing Russian economy. Since each unit costs around $90 million US and at least 21 vessels have been sold to foreign governments, including 10 units to India, 4 to China, 3 to Iran and 2 to Algeria, it suggests that almost $2 billion US has been raised through foreign export sales.

SPECIFICATIONS

Builder:	Sevmash
Class:	Kilo
Number:	B-459
Mission:	ASW submarine
Length:	72.6m (238ft)
Beam:	9.9m (32ft)
Displacement:	2450 tonnes (2489 tons)
Speed:	17 knots
Operating Depth:	240m (792ft)
Maximum Depth:	300m (990ft)
Crew:	53
Nuclear Weapons:	none
Conventional Weapons:	533mm (21in) torpedo
Sonar:	Rubikon sonar suite
Navigation:	GPS-based navigation system
Powerplant:	diesel-electric, 5900 shp
Date Commissioned:	unknown

YURIY DOLGORUKIY

The *Yuri Dolgorukiy* is a Borei class fourth-generation SSBN and was laid down at the Sevmash State Nuclear Shipbuilding Centre (shown above) at Sevmash in November 1996. The city of Moscow is sponsoring the project, as the lead vessel is named after Prince Dolgorukiy, the traditional founder of the city. So-called "presentation weapons" were commonplace in the Red Army during the Great Patriotic War. Presentation weapons were almost always the result of monetary collections taken up locally and voluntarily, and offered towards the cost of various vehicles or other items in the name of some personality or entity. Thus, the workers of a factory, town, or even just local citizens could take up a collection and buy a tank or aircraft in the name of their factory, group or a local figure. The Borei class will carry 20 SLBMs of a new type, yet the new SLBM has not yet been designed, which is a major reason for the delay in the completion of the first vessel. The intended missile, SS-N-28, failed its testing phase and was abandoned, and no replacement has been found. The lead unit of Russia's fourth-generation ballistic missile submarine would have reached initial operational capability by 2004, if the current plan of launching it by 2002 had remained on track. Suffice to say this has not happened, and the future of the project is unclear.

SPECIFICATIONS

Builder:	Sevmash
Class:	Borei
Number:	955
Mission:	ballistic missile submarine
Length:	170m (561ft)
Beam:	13.5m (45ft)
Displacement:	19,000 tonnes (19,400 tons)
Speed:	29 knots
Operating Depth:	unknown
Maximum Depth:	unknown
Crew:	110
Nuclear Weapons:	New SLBM
Conventional Weapons:	533mm (21in) torpedo
Sonar:	MGK-540 Skat-3M sonar suite
Navigation:	unknown
Powerplant:	nuclear reactors, 98,000 shp
Date Commissioned:	2005

ROKS LEE JONG MOO

The addition of the popular Type 209 German-designed attack submarine to the Republic of Korea Navy is part of an extensive expansion and modernization of its armed forces. The Changbogo class submarines, as the Type 209/1200 has been christened, are diesel-electric propulsion submarines, built under license in South Korea by the Daewoo company. ROKS *Changbogo*, the first ship of this class, was launched in June 1992 by HDW at their Kiel shipyards and commissioned in 1993. The second and subsequent boats were built by Daewoo Heavy Industries at Koje island, South Korea. The Changbogo class boats are far more sophisticated than any vessel in the North Korean fleet, and though North Korea has an advantage in terms of sheer numbers, they are old and often in poor condition. The covert incursion of North Korean submarines into South Korean territory on a frequent basis certainly keep the submariners of the ROK Navy busy and well-trained. According to the ROK naval doctrine, "The mission of the navy during peacetime is not only to deter war, but also to protect national and maritime sovereignty. Its mission during war is to guarantee the safety of our activities at sea by protecting the sea lines of communications, the life line of the country." The addition of the modern Changbogo class vessels goes a long way to fulfiling this strategic doctrine.

SPECIFICATIONS

Builder:	Daewoo Heavy Industries
Class:	Changbogo (Type 209/1200)
Number:	SS-066
Mission:	attack submarine
Length:	56m (187ft)
Beam:	6.2m (20ft)
Displacement:	1264 tonnes (1285 tons)
Speed:	22 knots
Operating Depth:	250m (825ft)
Maximum Depth:	300m (990ft)
Crew:	30
Nuclear Weapons:	none
Conventional Weapons:	533mm (21in) torpedo, mines
Sonar:	CSU 83 sonar suite
Navigation:	Raytheon SPS-10C radar
Powerplant:	diesel-electric, 5900 shp
Date Commissioned:	1998

NARVAL

The Delfin class patrol boat used by the Spanish Navy is a French-designed Daphne class vessel, but built under license in Spain with some modifications to the original design. This class of boat has been in operational service with the Spanish Navy for over 30 years, and is thus coming to the end of its operational life. Indeed, by some standards 30 years is an exceptionally long period of time for an attack submarine to remain in service given the advances in technology and computerization. In Spanish service, the Delfin class is scheduled to be replaced by the new Scorpène class very shortly, though as with many submarine-building projects around the world, the Scorpène class is behind schedule. Built in the 1950s through to the 1970s, the Daphne is a common attack submarine throughout the world. Of a simple conventional layout, the Daphne is a standard submarine, save the four aft torpedo tubes, which means that it has some measure of protection whilst retreating, or if it is caught unawares. However, this type of design shows the age of the boat and the era from which it came, since modern submarines do not generally incorporate a dedicated set of aft-facing torpedo tubes because they are considered redundant. The *Narval* represents the last of the original batch of Delfin submarines, and is being withdrawn from service in 2005.

SPECIFICATIONS

Builder:	ENB Cartagena Shipbuilding
Class:	Delfin (Daphne)
Number:	S-64
Mission:	attack/patrol submarine
Length:	57.6m (190ft)
Beam:	6.7m (22ft)
Displacement:	1043 tonnes (1060 tons)
Speed:	15 knots
Operating Depth:	200m (660ft)
Maximum Depth:	300m (990ft)
Crew:	56
Nuclear Weapons:	none
Conventional Weapons:	533mm (21in) torpedo, mines
Sonar:	Thomson Sintra DSUV 22A
Navigation:	unknown
Powerplant:	diesel-electric, 2000 shp
Date Commissioned:	1975

SCORPÈNE

The Spanish Navy has agreed to procure a number of Scorpène class attack submarines to replace its ageing Dolfin class boats. The class is derived from a family of advanced submarines designed by the French company DCN for export purposes, utilizing the technologies used in the French Navy's latest SSNs and SSBNs. There are three different types of Scorpène design, with the Spanish choice being the largest of the three which incorporates an Air Independent Propulsion (AIP) system. The Scorpène class is equipped with six bow-located 533mm (21in) torpedo tubes able to launch a variety of torpedoes, as well as surface-to-surface missiles. Up to 18 torpedoes and missiles can be carried, or 30 mines. The loading of weapons is automated, thus reducing the number of crew needed. The sonar suite includes a long-range passive cylindrical array, an intercept sonar, active sonar, distributed array, flank array, and a high-resolution sonar for mine and obstacle avoidance. The key planning concepts for the Scorpène class were to design an extremely quiet vessel with great detection capabilities and offensive power for missions ranging from anti-submarine and anti-surface warfare to special operations and intelligence gathering. Thus the Spanish Navy has invested in a submarine capable of providing it with excellent capabilities well into the twenty-first century.

SPECIFICATIONS

Builder:	Izar Cartagena Shipbuilding
Class:	S-80 Scorpène
Number:	S-80
Mission:	patrol/attack submarine
Length:	66.4m (219ft)
Beam:	6.2m (21ft)
Displacement:	1565 tonnes (1590 tons)
Speed:	20 knots
Operating Depth:	270m (891ft)
Maximum Depth:	350m (1155ft)
Crew:	33
Nuclear Weapons:	none
Conventional Weapons:	533mm (21in) torpedo, SSM
Sonar:	SUBTICS combat system
Navigation:	Kelvin-Hughes Type 1007
Powerplant:	diesel-electric AIP, 3800 shp
Date Commissioned:	2005

SIROCO

The Galerna coastal patrol submarine is a derivative of the French-designed and built Agosta class boat. As with all of the Spanish Navy's submarines, with the exception of the Scorpène class, the Galerna is a French submarine but built indigenously by the ENB Cartagena Shipbuilding Company, with improved electronic equipment. The original Agosta class was introduced into service in the 1970s, with an extensive refit taking place in the early 1990s to prolong the service life of the class, and make it more capable. This refit tailored the boats to fire the Exocet anti-shipping missile in addition to torpedoes. The Galerna class is used by the Spanish Navy primarily as a coastal patrol submarine, and thus it is not outfitted to be especially effective in deep-water environments. However, despite its limitations it is an effective weapons platform in littoral operations, and carries a formidable array of weapons for a vessel of its size. The Agosta class is currently in service with the French, Spanish and Pakistan navies. Four Galerna submarines were entered into service with the Spanish Navy during the mid-1980s, and all four remain on active duty. Their responsibilities include patrolling Spain's coastal areas, as well as its fishing territories. They can monitor the shipping coming in and out of Spanish waters, and challenge any vessels attempting to fish illegally or smuggle goods.

SPECIFICATIONS

Builder:	ENB Cartagena Shipbuilding
Class:	Galerna
Number:	S-72
Mission:	coastal patrol submarine
Length:	67.6m (223ft)
Beam:	6.8m (22ft)
Displacement:	1767 tonnes (1795 tons)
Speed:	20 knots
Operating Depth:	250m (825ft)
Maximum Depth:	300m (990ft)
Crew:	50
Nuclear Weapons:	none
Conventional Weapons:	533m (21in) torpedo, Exocet
Sonar:	DUUX-5
Navigation:	Thomson-CSF DRUA 33
Powerplant:	diesel-electric, 4600 shp
Date Commissioned:	1984

HMS NEPTUN

The Nacken class submarine is a fairly common diesel-electric boat, and is small and quiet, allowing for littoral warfare. As with the other diesel boats, the Nacken class is not an especially capable ocean-going vessel, but is instead designed for littoral operations. All the vessels of this class, a total of only two boats, have had their electronics and combat systems upgraded to the same standard as the Vastergotland class. The class is armed with eight torpedo tubes consisting of six 533mm (21in) tubes and two 400mm (15.7in) tubes. They can carry a total of 12 torpedoes, with 8 held in the tubes and the other 4 on the racks. The first-of-class vessel from which the class derives its name, HMS *Nacken,* is no longer in Swedish service. It was outfitted with the new Stirling Air Independent Propulsion (AIP) system and used as the test-bed for the next-generation Gotland class attack submarine. It was very successful in this role, and on the strength of its performance attracted admirers from other Scandinavian naval forces. After it was agreed that HMS *Nacken* could be sold with the AIP system, Denmark put in an offer for the boat. It was then subsequently sold to the Danish Navy with the AIP system intact, and was renamed *Kronborg*. The two remaining boats are still in Swedish service, but will be mothballed once the Gotland procurement is complete.

SPECIFICATIONS

Builder:	Kockums
Class:	Nacken
Number:	A-16
Mission:	patrol/attack submarine
Length:	49.5m (162ft)
Beam:	5.7m (19ft)
Displacement:	1145 tonnes (1163 tons)
Speed:	20 knots
Operating Depth:	250m (825ft)
Maximum Depth:	300m (990ft)
Crew:	19
Nuclear Weapons:	none
Conventional Weapons:	533mm (21in) torpedo, mines
Sonar:	Atlas CSU-83 sonar suite
Navigation:	Terma radar
Powerplant:	diesel-electric, 1500 shp
Date Commissioned:	1981

HMS OSTERGOTLAND

The Sodermanland class attack submarine is an improved version of the original Vastergotland class submarine that has been the Swedish Navy's premier attack vessel since the late 1980s. Despite its relatively small size it is a capable weapons platform. In terms of technology the Sodermanland is easily outclassed by the new Gotland boats, but it will remain in service. This is due to the fact that the final two boats of the Vastergotland class will be fitted with the Stirling Air Independent Propulsion(AIP) system that is fitted to the Gotland class, and renamed the Sodermanland class. The conversion is an extensive operation. The submarines will be cut in two immediately aft of the tower and lengthened by the insertion of the Stirling AIP section. This section, fully fitted and equipped before installation, contains two Stirling units, liquid oxygen (LOX) tanks and electrical equipment. The inclusion of this section increases the length by almost 12m (40ft). The first boat to be converted is the *Sodermanland*, which will be relaunched in 2003, and the *Ostermanland* will be relaunched in 2004. Another aspect of the class conversion is that the submarines will be equipped to undertake international peacekeeping missions in warmer and more saline waters. This conversion includes the addition of different filtration systems and heating/cooling apparatus.

SPECIFICATIONS

Builder:	Kockums
Class:	Sodermanland
Number:	unknown
Mission:	patrol/attack submarine
Length:	60.5m (200ft)
Beam:	6.1m (20ft)
Displacement:	1500 tonnes (1524 tons)
Speed:	20 knots
Operating Depth:	300m (990ft)
Maximum Depth:	400m (1320ft)
Crew:	30
Nuclear Weapons:	none
Conventional Weapons:	533mm (21in) torpedo, SSM
Sonar:	Atlas Electronik CSU83 sonar suite
Navigation:	Terma radar
Powerplant:	diesel-electric, 1800 shp
Date Commissioned:	1990 (relaunched 2004)

HMS UPPLAND

The Gotland class attack submarine is the latest addition to the Royal Swedish Navy's submarine fleet, having been in service since 1997. This class is essentially an improved version of its predecessor, the Vastergotland class, and is one of the world's most modern conventionally powered submarines, and one of the finest. It is capable of achieving a variety of missions, such as anti-shipping operations, anti-submarine missions, forward surveillance, special operations and minelaying. During wartime, the Gotland class would be employed to gather intelligence on the enemy, lay mines close to the coast to deter invasion, and harass enemy vessels. To achieve this, these submarines can carry a powerful array of wire-guided and homing torpedoes, missiles and mines. Saab Bofors Underwater Systems has developed a new heavyweight torpedo for the Swedish Navy, the Torpedo 2000, built with the Gotland submarines in mind. It is a high-speed anti-submarine/anti-surface torpedo with a range of more than 40km (25 miles) and a speed of over 40 knots. The Gotland class was also the world's first conventional submarine originally designed with an Air Independent Propulsion (AIP) system. Other conventional submarines have begun to incorporate this cutting-edge technology, but the Gotland was the first to enter active service.

SPECIFICATIONS

Builder:	Kockums HDW
Class:	Gotland
Number:	A-20
Mission:	attack submarine
Length:	60m (199ft)
Beam:	6.2m (20ft)
Displacement:	1500 tonnes (1524 tons)
Speed:	20 knots
Operating Depth:	300m (990ft)
Maximum Depth:	450m (1485ft)
Crew:	25
Nuclear Weapons:	none
Conventional Weapons:	533mm (21in) torpedo, Harpoon
Sonar:	Atlas Electronik CSU 90-2 sonar
Navigation:	Terma Scanter navigation radar
Powerplant:	diesel-electric AIP
Date Commissioned:	1997

HMS VASTERGOTLAND

The Vastergotland class attack submarine has been the Swedish Navy's primary "hunter-killer" submarine since the late 1980s. Despite having been left behind technologically by the new Gotland class and improved Sodermanland class boats, it is still a capable weapons platform. The class is fitted with nine torpedo tubes; six bow mounted 533mm (21in) tubes, and three aft mounted 400mm (15.7in) tubes. They can launch two different calibres of torpedo, the type 613 heavyweight torpedo, and the type 431/451 lightweight torpedo. It can also place mines in lieu of carrying torpedoes. The Vastergotland class operates almost exclusively in the cold waters of the North Sea and northern Atlantic, as well as in the Baltic Sea. Though they are not especially old vessels, they will be placed in reserve once all the Gotland and Sodermanland class boats have entered service. The advanced capabilities of these two classes means that the original Vastergotland class is unable to match them operationally, and thus the cost of maintaining them is not warranted. If a state of war was to break out, however, the two mothballed vessels would be able to be brought back into service within a very short space of time. Whether there would be the manpower and expertise needed to operate them effectively is debatable. In reality, though, there is little chance they will be needed.

SPECIFICATIONS

Builder:	Kockums
Class:	Vastergotland
Number:	A-17
Mission:	patrol/attack submarine
Length:	48.5m (160ft)
Beam:	6.1m (20ft)
Displacement:	1143 tonnes (1161 tons)
Speed:	20 knots
Operating Depth:	300m (990ft)
Maximum Depth:	400m (1320ft)
Crew:	30
Nuclear Weapons:	none
Conventional Weapons:	533mm (21in) torpedo, Harpoon
Sonar:	Atlas Electronik CSU-83 sonar
Navigation:	Terma radar
Powerplant:	diesel-electric, 1800 shp
Date Commissioned:	1988

HAI HU

The story of Taiwanese attempts to procure or produce good-quality conventionally powered submarines is one of the ongoing sagas of the international arms trade. Taiwan has long been in the market for additional diesel submarines to counter China's growing naval might. Under its proposed submarine programme, Taiwan's navy plans to increase its submarine fleet to 12 vessels. But Taiwan's desires to add new submarines to the navy have tended to move no further than paper agreements. The main submarine exporters, such as Germany and the United States, have not endorsed Taiwan's proposals because of fears it could upset the balance between the Chinese mainland and Taiwan, and provoke an international crisis. This fear is not without grounds, since China downgraded its diplomatic relations with the Netherlands following the sale of the two Zwaardvis boats to Taiwan in the late 1980s. Thus Taiwan's navy currently has only four submarines, two of which are too old for operations and are used only as training vessels. The remaining two are the Dutch vessels. During hostilities, these two Hai Lung class submarines could be used to protect Taiwan's shipping lanes and ports, preventing them from being mined, or could even take on offensive operations and attack enemy shipping and mine the enemy's ports with a compliment of torpedoes, mines and missiles.

SPECIFICATIONS

Builder:	Wilton Fijenoord
Class:	Hai Lung (Zwaardvis)
Number:	794
Mission:	attack/patrol submarine
Length:	66.9m (219ft)
Beam:	8.4m (27ft)
Displacement:	2660 tonnes (2702 tons)
Speed:	20 knots
Operating Depth:	240m (792ft)
Maximum Depth:	300m (990ft)
Crew:	67
Nuclear Weapons:	none
Conventional Weapons:	533mm (21in) torpedo, SSM
Sonar:	Signaal SIASS-Z sonar suite
Navigation:	Signaal ZW-07
Powerplant:	diesel-electric, 5100 shp
Date Commissioned:	1988

TCG ANAFARTALAR

The Preveze class attack submarine is the Turkish Navy's version of the German-designed Type 209 submarine from the well-known ship designers HDW. It is also the latest addition to their submarine fleet. The Anafartalar is the enlarged Type 209/1400 version, which displaces slightly more than its Atilay class predecessor. The Turkish Navy has ordered an additional number of these boats to supplement an already impressive submarine force. Though the Turkish Navy still operates some ancient World War II Guppy class submarines, the new Type 209 Preveze class will make its fleet more modern and enhance its military capabilities many times. This class of boat is designed for coastal patrol and attack missions in littoral environments, rather than blue-water operations. The chief theatre of operations for the Turkish Navy is in and around the Mediterranean, with occasional expeditions into the Atlantic and elsewhere. A prime task of the submarine force is to secure Turkey's vital shipping lanes, since the country depends heavily on its freedom to import and export goods via giant cargo ships without fear of sabotage. Furthermore, Turkish submarines take part in the constant ritual of spying on its traditional enemy, Greece. Aside from monitoring Greek naval manoeuvres, the Type 209 vessels continue the vigil on the divided island of Cyprus.

SPECIFICATIONS

Builder:	Gölcük Naval Yard
Class:	Preveze (Type 209/1400)
Number:	S-365
Mission:	attack submarine
Length:	62m (204ft)
Beam:	6.2m (20ft)
Displacement:	1586 tonnes (1611 tons)
Speed:	21 knots
Operating Depth:	300m (990ft)
Maximum Depth:	350m (1155ft)
Crew:	30
Nuclear Weapons:	none
Conventional Weapons:	533mm (21in) torpedo, Harpoon
Sonar:	CSU-81/1 sonar suite
Navigation:	unknown
Powerplant:	diesel-electric, 5000 shp
Date Commissioned:	1999

TCG DOLUNAY

The Turkish Navy's Atilay class attack submarine is a close relation to its Preveze class cousin. The class, of which TCG *Dolunay* is one, is an older Type 209/1200 submarine from the same German designers that delivered the Preveze class, Howaldswerke-Deutsche Werft (HDW). The introduction of the original Type 209s in the mid-1970s signalled a new era for the Turkish submarine fleet. The contract between the German shipbuilders and the Turkish Navy agreed that the first batch of three submarines would be built at Kiel by HDW, whilst the remaining boats would be built with German assistance at the Gölcük shipyards in Turkey. The type proved so successful that the Preveze class was ordered on the strength of the Atilay's performance, and the relationship between the Turkish Navy and HDW has seen a second batch of Type 209/1400 Preveze class ordered. The generic Type 209/1200 is designed as a coastal submarine with anti-submarine and anti-surface ship warfare in mind, along with the protection of naval bases, coastal installations and sea lanes, and also for general reconnaissance and patrol missions. The vessel is armed with eight 533mm (21in) torpedo tubes, and is able to fire the AEG SST-4 torpedo, of which it carries 14, as well as other types of mines and countermeasures. It cannot fire surface-to-surface missiles however.

SPECIFICATIONS

Builder:	Gölcük Naval Yard
Class:	Atilay (Type 209/1200)
Number:	S-352
Mission:	attack submarine
Length:	55.9m (183ft)
Beam:	6.3m (21ft)
Displacement:	1285 tonnes (1306 tons)
Speed:	22 knots
Operating Depth:	250m (825ft)
Maximum Depth:	350m (1155ft)
Crew:	33
Nuclear Weapons:	none
Conventional Weapons:	533mm (21in) torpedo, mines
Sonar:	CSU-3 sonar suite
Navigation:	unknown
Powerplant:	diesel-electric, 5000 shp
Date Commissioned:	1990

ASTUTE

The new Astute class submarine is the United Kingdom's latest nuclear powered hunter-killer attack submarine, designed to replace the ageing Swiftsure class. Though based on the Trafalgar class, indeed they were originally known as Batch 2 Trafalgar, they are a major upgrade with an extensively modified front hull. When the Astute class comes into operation in 2006, it will be one of the most capable submarines anywhere in the world. As with all first-rate military powers, the United Kingdom has had to adapt its military doctrine to embrace the new concepts of warfare. Thus the Astute class has been designed with flexibility in mind, able to operate in many different environments at short notice, and use its full compliment of sophisticated weaponry and systems to aid any particular task force. There exists an order for the first batch of three Astute class submarines to be delivered in the first decade of the twenty-first century, and BAE Systems Marine, the prime contractor, expects another order of three to be placed. The Astute is equipped with the Tomahawk Block III cruise missile (TLAM) from US company Raytheon and the Sub Harpoon anti-shipping missile produced by Boeing. Both are fired from 533mm (21in) torpedo tubes. Also carried is the Spearfish torpedo, which has a range is 65km (40 miles).

SPECIFICATIONS

Builder:	BAE Systems Marine
Class:	Astute
Number:	unknown
Mission:	attack submarine
Length:	91.7m (303ft)
Beam:	10.4m (34ft)
Displacement:	7200 tonnes (7315 tons)
Speed:	29 knots
Operating Depth:	300m (990ft)
Maximum Depth:	400m (1320ft)
Crew:	110
Nuclear Weapons:	none
Conventional Weapons:	533mm (21in) torpedo, TLAM
Sonar:	Thales 2076 sonar suite
Navigation:	I-band radar suite
Powerplant:	nuclear reactor, 15,000 shp
Date Commissioned:	2006

HMS SOVEREIGN

The ageing Swiftsure class attack submarine was built and added to the Royal Navy's submarine fleet to supplement existing fleet submarines and build up nuclear submarine numbers at a time when the United Kingdom only had a few. The Swiftsure class was a follow-on from the successful Valiant Class attack submarine. However, a number of improvements were incorporated into the new design including a cylindrical hull and improved sonar and torpedo systems. As a result of these improvements, the Swiftsure class was a marked step forward, as they were quieter, faster and could dive to greater depths than any previous British SSN, but they are now lagging far behind in terms of technology and capability. As with all attack submarines, the principle role of these "hunter-killer" vessels is to attack ships and other submarines. In this capacity they could support and protect a convoy or taskforce. Additionally fleet submarines can be used in a surveillance role. The Swiftsure was a surprise choice to be the first British submarine to be fitted with the ability to fire the Tomahawk cruise missile, and has thus acquired a land attack role as well. The Swiftsure class is approaching the end of its operational life, having had a varied and active service, but is due to be replaced by the new Astute class SSN when it enters service in 2006.

SPECIFICATIONS

Builder:	Vickers Shipbuilding
Class:	Swiftsure
Number:	S-108
Mission:	attack submarine
Length:	82.9m (273ft)
Beam:	9.8m (32ft)
Displacement:	5000 tonnes (5080 tons)
Speed:	30 knots
Operating Depth:	230m (760ft)
Maximum Depth:	300m (990ft)
Crew:	116
Nuclear Weapons:	none
Conventional Weapons:	533mm (21in) torpedo, Harpoon
Sonar:	Marconi 2074 sonar suite
Navigation:	Kelvin Hughes Type 1006
Powerplant:	nuclear reactor, 15,000 shp
Date Commissioned:	1974

HMS SPLENDID

The Royal Navy Swiftsure class attack submarine HMS *Splendid* has been one of the most active vessels in the United Kingdom's submarine fleet over the last 20 years. She joined the fleet in 1981, and saw immediate action when the Argentine military junta invaded the Falkland Islands in 1982. She was part of the Task Force that set sail to reconquer the islands, and was one of four submarines that took part in the conflict. Though the submarine HMS *Conqueror* grabbed the headlines by sinking the Argentine cruiser *General Belgrano*, the presence of the four submarines persuaded the Argentine Navy to stay out of the war. After being retro-fitted in 1998, HMS *Splendid* became the first Royal Navy submarine capable of firing Tomahawk cruise missiles. She demonstrated this new ability when she destroyed a small building in the United States, 640km (400 miles) inland. When the strikes against Serbia began in March 1999 during the Kosovo crisis, HMS *Splendid* joined US warships in launching missiles at Serb targets, becoming the first British nuclear submarine to fire in anger since HMS *Conqueror* in 1982, reportedly destroying a radar site at Pristina airport. Though in the grand scheme of things this was more a gesture than any great military importance, it signified a step forward in the Royal Navy's fighting capabilities.

SPECIFICATIONS

Builder:	Vickers Shipbuilding
Class:	Swiftsure
Number:	S-112
Mission:	attack submarine
Length:	82.9m (273ft)
Beam:	9.8m (32ft)
Displacement:	5000 tonnes (5080 tons)
Speed:	30 knots
Operating Depth:	230m (760ft)
Maximum Depth:	300m (990ft)
Crew:	116
Nuclear Weapons:	none
Conventional Weapons:	533mm (21in) torpedo, TLAM
Sonar:	Marconi 2074 sonar suite
Navigation:	Kelvin Hughes Type 1006
Powerplant:	nuclear reactor, 15,000 shp
Date Commissioned:	1981

HMS TRAFALGAR

The Trafalgar class attack submarine was the United Kingdom's primary anti-submarine and anti-shipping vessel during the latter years of the Cold War and into the 1990s. It was originally designed for Cold War operations in the Mediterranean and North Atlantic theatres, but has seen service in other parts of the world. The design is a follow-on from the successful Swiftsure class attack submarine but incorporated many improvements, making the class both faster and quieter than any previous British nuclear submarine. The principle role of these hunter-killer vessels, as they are colloquially known in the British submarine service, is to attack ships and other submarines that might otherwise endanger a convoy or task force. To achieve this aim the Trafalgar class is fitted with five 533mm (21in) torpedo tubes, which can fire either the Spearfish or Tigerfish torpedoes, launch the Sub Harpoon missile or deploy mines. Two of the class, HMS *Triumph* and HMS *Trafalgar*, have been retro-fitted to fire Tomahawk cruise missiles. It is expected that the remaining five boats of the class will all be able to launch the Tomahawk missile by 2006. The Trafalgar class has recently been under the media spotlight for a series of faults and cracks that forced all seven boats to be grounded for a short while. Despite this setback, the Trafalgar class submarine is still a most capable vessel.

SPECIFICATIONS

Builder:	Vickers Shipbuilding
Class:	Trafalgar
Number:	S-107
Mission:	attack submarine
Length:	85.3m (281ft)
Beam:	9.8m (32ft)
Displacement:	5200 tonnes (5283 tons)
Speed:	30 knots
Operating Depth:	300m (990ft)
Maximum Depth:	400m (1320ft)
Crew:	130
Nuclear Weapons:	none
Conventional Weapons:	533mm (21in) torpedo, TLAM
Sonar:	BAe Type 2007 sonar suite
Navigation:	Kelvin Hughes Type 1007
Powerplant:	nuclear reactor, 15,000 shp
Date Commissioned:	1983

HMS TRIUMPH

HMS *Triumph* is the last in a batch of seven Trafalgar class attack submarines to enter service with the British Royal Navy. Aside from the traditional "hunter-killer" role of the Trafalgar class, these boats are able to conduct intelligence-gathering operations and perform a surveillance role. These operations might involve moving close to enemy forces at sea, and monitoring their operations and movements whilst remaining undetected. This type of surveillance may also include underwater photography, sometimes of surface vessels. Similarly, the surveillance role may include the monitoring of a stretch of coastline using sophisticated video technology or digital photography. A submarine could stealthily approach a coastline in shallow water and assess the situation prior to an amphibious invasion or land action. To achieve this, all Trafalgar class boats are fitted with camera equipment and thermal-imaging scopes in their periscope arrays. The ability of an SSN to remain on active duty, totally independent from the need for support vessels and with flexible mission tasks, makes them useful in modern warfare. This independence was illustrated by HMS *Triumph* when she sailed to Australia in 1993 travelling 65,600km (41,000) miles submerged without any forward support. This remains the longest ever solo deployment by a nuclear submarine.

SPECIFICATIONS

Builder:	Vickers Shipbuilding
Class:	Trafalgar
Number:	S-93
Mission:	attack submarine
Length:	85.3m (281ft)
Beam:	9.8m (32ft)
Displacement:	5200 tonnes (5283 tons)
Speed:	30+ knots
Operating Depth:	300m (990ft)
Maximum Depth:	400m (1320ft)
Crew:	130
Nuclear Weapons:	none
Conventional Weapons:	533mm (21in) torpedo, TLAM
Sonar:	BAe Type 2007 sonar suite
Navigation:	Kelvin Hughes Type 1007
Powerplant:	nuclear reactor, 15,000 shp
Date Commissioned:	1991

HMS VANGUARD

The new Vanguard Class SSBN (Ship Submersible Ballistic Nuclear) provides the United Kingdom's only strategic and sub-strategic nuclear deterrent. The first Vanguard class submarine was launched in 1993 and was the replacement for the ageing Polaris SSBNs of the Resolution class. This class of submarines is extremely capable and provides the United Kingdom with a wide range of capabilities far beyond that of simple strategic deterrence. In its role of SSBN, the Vanguard has the capacity to carry 16 D-5 Trident II missiles, each with up to 12 independently targetable nuclear warheads. One Vanguard class boat can carry up to 192 individual warheads, though in reality it carries a maximum of 96, which has been recently reduced to 48. The Trident II missile has a range of up to 11,000km (6875 miles), and each warhead can hit its target with an accuracy of within 120m (360ft). In addition to its capability to launch nuclear missiles, the Vanguard class has been recently configured to launch the Tomahawk cruise missile (TLAM), which gives the vessel greater flexibility and capability. Each vessel of this class has four 533mm (21in) tubes, carrying the Tigerfish and Spearfish torpedoes. The Tigerfish Mark 24 Mod 2 torpedo is a wire-guided torpedo with a 134kg (295lb) warhead and the Spearfish is a wire-guided torpedo with a range of 65km (40 miles).

SPECIFICATIONS

Builder:	Vickers Shipbuilding
Class:	Vanguard
Number:	S-28
Mission:	ballistic missile submarine
Length:	149.3m (493ft)
Beam:	12.8m (42ft)
Displacement:	16,000 tonnes (16,256 tons)
Speed:	25 knots
Operating Depth:	300m (990ft)
Maximum Depth:	450m (1485ft)
Crew:	132
Nuclear Weapons:	16 x Trident II D-5 SLBMs
Conventional Weapons:	533mm (21in) torpedo, TLAM
Sonar:	BAE 2054 composite sonar system
Navigation:	Type 1007 I-band radar
Powerplant:	nuclear reactor, 27,500 shp
Date Commissioned:	1994

HMS VICTORIOUS

HMS *Victorious* is another one of the Royal Navy's four Vanguard class SSBNs, and was commissioned into service in 1995. In the face of decreasing defence budgets and changing geopolitical climates, the Vanguard class has been able to adapt itself from a purely strategic role, to something altogether more flexible, and thus ultimately more worthy of its phenomenal cost. The Vanguard class can now launch the Tomahawk cruise missile, which is able to strike at inland targets at ranges of over 1300km (400 miles) with great accuracy. This means that the Vanguard class can deploy covertly to any troublespot in the world, and sit off the coast collecting intelligence, able to strike at any moment should the order be given. The extremely powerful intelligence gathering capabilities of the Vanguard class, along with its state-of-the-art communications arrays, is also one of its most useful assets. During the conflict in the Balkans throughout the 1990s, and in the more recent Kosovo crisis, all communications traffic for the entire NATO alliance was sent through a Vanguard class submarine hiding in the Adriatic. Some have questioned the continuing utility of a strategic nuclear deterrent, but the Vanguard has shown itself able to cope with the rigours of the post-Cold War world, and able to find a place for its capabilities within the United Kingdom's force structure.

SPECIFICATIONS

Builder:	Vickers Shipbuilding
Class:	Vanguard
Number:	S-29
Mission:	ballistic missile submarine
Length:	149.3m (493ft)
Beam:	12.8m (42ft)
Displacement:	16,000 tonnes (16,293 tons)
Speed:	25 knots
Operating Depth:	300m (990ft)
Maximum Depth:	450m (1485ft)
Crew:	132
Nuclear Weapons:	16 x Trident II D-5 SLBMs
Conventional Weapons:	533mm (21in) torpedo, TLAM
Sonar:	BAE 2054 composite sonar system
Navigation:	Type 1007 I-band radar
Powerplant:	nuclear reactor, 27,500 shp
Date Commissioned:	1995

ASDS

Submarines have long been used for special operations carrying commandos, reconnaissance teams and agents on high-risk missions. Most special operations in US submarines are carried out by SEALs, the Sea-Air-Land teams trained for missions behind enemy lines. These special forces can be inserted by fixed-wing aircraft, helicopter, parachute or surface craft, but in most scenarios only submarines guarantee covert delivery. Once in the objective area, SEALs can carry out reconnaissance, monitoring enemy movements, and a host of other clandestine and often dangerous missions. Submarines are especially well suited for this role because of their high speed, endurance and stealth. The Advanced SEAL Delivery System (ASDS) is a long-range miniature submersible capable of delivering special forces for covert missions. ASDS provides improved range, speed and payload, and habitability for the crew and a SEAL squad. ASDS will be carried to its operational area by a host ship, currently a specially configured Los Angeles class, and in the future a Seawolf or Virginia class submarine. ASDS will also be air transportable by either C-5 or C-17 aircraft. A total of six ASDS will be built for the US Special Operations Command (USSOCOM), and will be able to perform a variety of tasks including covert insertion, reconnaissance and rescue.

SPECIFICATIONS

Builder:	Northrup Grumman
Class:	ASDS
Number:	n/a
Mission:	SEAL delivery submarine
Length:	19.8m (64ft)
Beam:	2.4m (8ft)
Displacement:	60 tonnes (59 tons)
Speed:	8 knots
Operating Depth:	unknown
Maximum Depth:	unknown
Crew:	2 + 8 SEALs
Nuclear Weapons:	none
Conventional Weapons:	none
Sonar:	classified
Navigation:	classified
Powerplant:	electric motor, 67 shp
Date Commissioned:	2001

UNITED STATES

EX-USS TROUT

The last operational diesel-electric submarine in the US Navy has been designated and modified for use as an operational underwater sonar target for anti-submarine warfare (ASW) exercises. The vessel will be operated by Naval Air Warfare Center Aircraft Division (NAWCAD) Detachment Key West. In addition to ASW training, the boat is used for testing and trials of new technologies. It is manned by a small crew while surfaced for transit, but unmanned while submerged. The boat is not assigned an official designation or name. Formerly USS *Trout*, it was decommissioned and stricken on 19 December 1978 and nominally transferred to Iran when relations between the two countries were buoyant. However, the vessel was never delivered following the Iranian revolution and she was laid up at Philadelphia while her ownership and fate were worked out. In 1994 she was sold back to the US Navy at scrap value. A vital training tool for the newest generation of anti-submarine units, the ex-USS *Trout* allows the US Navy to train its airborne naval anti-submarine units, such as the SH-60B Seahawk helicopter crews, to the highest level. It also gives them practical experience of hunting an actual submarine that has similar capabilities to other potential enemy vessels around the world, such as those used by China, North Korea and other Middle East states.

SPECIFICATIONS

Builder:	Electric Boat
Class:	SSN-563 Tang (Modified)
Number:	SSN-566
Mission:	ASW training submarine
Length:	84.7m (278ft)
Beam:	8.2m (27ft)
Displacement:	2700 tonnes (2743 tons)
Speed:	16 knots
Operating Depth:	unknown
Maximum Depth:	unknown
Crew:	0 + 10
Nuclear Weapons:	none
Conventional Weapons:	none
Sonar:	unknown
Navigation:	unknown
Powerplant:	diesel-electric, 5600 shp
Date Commissioned:	1951 (re-introduced 1994)

NR-1

NR-1, the first deep submergence vessel using nuclear power, was launched on 25 January 1969, and successfully completed her initial sea trials on 19 August 1969. It manoeuvres by four ducted thrusters, two in the front and two in the rear. The vehicle also has planes mounted on the sail, and a conventional rudder. NR-1's official missions have included search, object recovery, geological survey, oceanographic research, and installation and maintenance of underwater equipment. It was also in action following the loss of the Space Shuttle Challenger in 1986, searching for, identifying, and recovering critical parts of the Challenger craft. Militarily the NR-1 can be used for a number of covert operations in very deep water. It is doubtful that special forces would be deployed via the NR-1, but the vessel could be used for reconnaissance, for the emplacement of seabed acoustic devices and even the covert emplacement of mines in heavily defended areas. Though the NR-1 is not officially a military asset and has been used in a large number of civilian environments, such as archaeological studies, the vessel is certainly used for covert operations. The title "research vessel" is an often-used military euphemism for vessels used in clandestine missions, and the NR-1 is no exception. Despite its relative age, it is still a useful tool.

SPECIFICATIONS

Builder:	Electric Boat
Class:	NR-1
Number:	NR-1
Mission:	research/special operations
Length:	41.5m (136ft)
Beam:	3.6m (12ft)
Displacement:	394 tonnes (387 tons)
Speed:	5 knots
Operating Depth:	n/a
Maximum Depth:	1000m (3300ft)
Crew:	13
Nuclear Weapons:	none
Conventional Weapons:	none
Sonar:	Nautronics ATS
Navigation:	Benthos TR6000
Powerplant:	nuclear reactor
Date Commissioned:	1969

USS CHEYENNE

The USS *Cheyenne* is the United States Navy's most modern Los Angeles class attack submarine. It is 20 years younger than the original Los Angeles class boat, and thus it has incorporated some of military technology's more recent innovations into the same basic design. Though sharing many of the external appearances of the original Los Angeles class vessels, internally the newest boats are almost a different class given the vast improvements in computer technology during the 20 years that the Los Angeles class has been in production. The final 23 hulls (SSN-751 and later), referred to as 688I, are quieter, incorporate an advanced BSY-1 sonar suite combat system and have the ability to lay mines from their torpedo tubes. The weapons systems onboard a Los Angeles class submarine are truly awesome. The attack centre is the operational heart of the ship, and from where all offensive and defensive orders are made. The integrated fire-control computer is able to detect and classify targets and prepare a firing solution extremely quickly. The Mk48 torpedo that is used against other submarines or surface vessels is capable of travelling up to 55 knots and hitting a target over 8km (5 miles) away. Additionally, the Improved Los Angeles class has been configured for under-ice operations with the inclusion of a strengthened sail and bow planes moved from the sail.

SPECIFICATIONS

Builder:	Newport News
Class:	SSN-688 Improved Los Angeles
Number:	SSN-773
Mission:	attack submarine
Length:	109.7m (360ft)
Beam:	10m (33ft)
Displacement:	6210 tonnes (6845 tons)
Speed:	32 knots
Operating Depth:	300m (950ft)
Maximum Depth:	400m (1320ft)
Crew:	129
Nuclear Weapons:	none
Conventional Weapons:	533mm (21in) torpedo, TLAM
Sonar:	BSY-1 sonar suite
Navigation:	AN/BPS-15 navigation radar
Powerplant:	nuclear reactor, 35,000 shp
Date Commissioned:	1996

USS CHICAGO

The USS *Chicago*, though 10 years younger than her sister ship USS *Los Angeles*, is an almost identical vessel. Both boats use the same technology for fire control, navigation, sonar and propulsion. The S6G nuclear reactor on the Los Angeles class is a modified version of the D2G reactor first used on the CGN-25 Bainbridge class of guided missile cruiser. The nuclear powerplant gives these boats the ability to remain deployed and submerged for long periods of time, up to months at a time. Whilst the machinery is able to operate at sea for such long periods, the crew certainly could not without sophisticated life-support systems. Thus, the ships are outfitted with auxiliary equipment to provide for the needs of the crew. Atmosphere-control equipment replenishes oxygen used by the crew, and removes CO_2 and other atmospheric contaminants. The ship is equipped with two distilling plants, able to convert thousands of gallons of salt water into fresh water for drinking, washing and the propulsion plant. Sustained operation of the complex equipment and machinery on the ship requires the support of repair parts carried on board. The ship carries enough food to feed the crew for as long as 90 days. The logistical effort needed to keep a Los Angeles class attack submarine in fully operational order is a massive task requiring the expertise of many sailors.

SPECIFICATIONS

Builder:	Newport News
Class:	SSN-688 Los Angeles
Number:	SSN-721
Mission:	attack submarine
Length:	109.7m (360ft)
Beam:	10m (33ft)
Displacement:	6210 tonnes (6845 tons)
Speed:	32 knots
Operating Depth:	300m (950ft)
Maximum Depth:	450m (1475ft)
Crew:	129
Nuclear Weapons:	none
Conventional Weapons:	533mm (21in) torpedo, TLAM
Sonar:	AN/BQQ-5D sonar array
Navigation:	AN/BPS-15 navigation radar
Powerplant:	nuclear reactor, 35,000 shp
Date Commissioned:	1986

UNITED STATES

USS CONNECTICUT

The USS *Connecticut* is the second boat in the Seawolf series, and to all intents and purposes is identical to the USS *Seawolf*. Whilst the Seawolf class of submarine is a technological masterpiece in many respects, it was designed simply to counter the threat posed by Soviet submarines. However, the fall of the Soviet Union robbed the Seawolf class of its intended prey, and thus in the changed geopolitical climate, and given the shift in the manner in which wars are fought, USS *Connecticut* is searching for a role. The massive costs of the Seawolf programme made the submarine, at $3.5 billion US, prohibitively expensive and the number of submarines was limited to just three. However, all was not lost because much of the research effort that went into Seawolf has been used to good effect in the design of Seawolf's cheaper successor, the Virginia class SSN. Nonetheless, the USS *Connecticut* is a formidable machine, capable of taking on targets both on land, on the sea and under the waves more or less simultaneously. The Seawolf class can carry more armaments in a more varied package than any previous US SSN. It can hold 50 Tomahawk missiles, or 50 Sub Harpoon missiles, or 50 Mk48 torpedoes, or up to 100 mines. The mixture of armaments would be dependent on the mission, but a combination of all four types would be carried.

SPECIFICATIONS

Builder:	General Dynamics Electric Boat
Class:	SSN-21 Seawolf
Number:	SSN-22
Mission:	attack submarine
Length:	107.6m (353 ft)
Beam:	12.2m (40 feet)
Displacement:	9137 tonnes (9283 tons)
Speed:	35 knots
Operating Depth:	400m (1320ft)
Maximum Depth:	classified
Crew:	133
Nuclear Weapons:	none
Conventional Weapons:	660mm (30in) torpedo, TLAM
Sonar:	BSY-2 sonar/combat suite
Navigation:	BPS-16 navigation radar
Powerplant:	nuclear reactor, 52,000 shp
Date Commissioned:	1998

USS JIMMY CARTER

The USS *Jimmy Carter* is the third and final boat in the Seawolf class of submarines. However, Department of Defense chiefs have used the cancellation of the Seawolf series as an opportunity to use the advanced technology of the Seawolf in a novel and modified way, and the USS *Jimmy Carter* is the boat they have selected to experiment with. This said, the USS *Jimmy Carter* will be fully operational, but will utilize new technologies and techniques to explore the potential future of submarine warfare now that the traditional Soviet threat has diminished and the nature of modern conflict appears to require a different doctrine for submarines. To this end the USS *Jimmy Carter* has been comprehensively modified. The planned alterations include lengthening the hull section behind the sail and inserting an Ocean Interface section that will support a new Multi-Mission Project by opening larger payload apertures to the sea. It will also be able to support future concepts of offensive and defensive mine warfare in her ability to launch and recover a wide range of tethered and autonomous vehicles and sensors of varying sizes and shapes. In addition to these robust capabilities, USS *Jimmy Carter* will also be capable of supporting special operations forces with provision for operating the Dry Deck Shelter (DDS) and Advanced SEAL Delivery System (ASDS).

SPECIFICATIONS

Builder:	Electric Boat
Class:	SSN-21 Seawolf
Number:	SSN-23
Mission:	special operations submarine
Length:	107.6m (353 ft)
Beam:	12.2m (40 feet)
Displacement:	9137 tonnes (9283 tons)
Speed:	35 knots
Operating Depth:	400m (1325ft)
Maximum Depth:	classified
Crew:	133 + 50 special forces
Nuclear Weapons:	none
Conventional Weapons:	660mm (30in) torpedo, TLAM
Sonar:	BSY-2 sonar / combat suite
Navigation:	BPS-16 navigation radar
Powerplant:	nuclear reactor, 52,000 shp
Date Commissioned:	2001

USS KAMEHAMEHA

Originally commissioned as ballistic missile submarines, two remaining members of the Benjamin Franklin class of submarine were converted to special operations attack submarines, with a capacity for carrying and delivering special operations forces. They are equipped to covertly insert special operations forces into hostile territory. Like all submarines, they can sit off a coast for as long as needed, undetected, waiting to act or leave without raising tensions in the meantime. The boomers converted to dual Dry Dock Shelter (DDS) carriers are huge compared to the Sturgeon SSNs previously used for this role. On the SSBN the SEALs and crew can be berthed in greater comfort. Enough exercise equipment can be loaded for the SEALs and crew to maintain the physical conditioning required for mission success. SEAL mission planning, briefings and operations can be conducted with minimum crew disruption, whilst the crew are free to operate as if the SEALs were not on board. In August 1993, USS *Kamehameha* arrived in Pearl Harbor to become part of Submarine Squadron One. This vessel now regularly deploys in support of special warfare objectives throughout the Pacific and beyond. The USS *Kamehameha* is a very capable attack and special operations submarine, but is approaching the end of its service life.

SPECIFICATIONS

Builder:	Electric Boat
Class:	Modified Benjamin Franklin
Number:	SSN-642
Mission:	attack/SEAL delivery submarine
Length:	129m (425ft)
Beam:	10m (33ft)
Displacement:	8120 tonnes (8250 tons)
Speed:	25 knots
Operating Depth:	300m (990ft)
Maximum Depth:	400m (1320ft)
Crew:	135 + 65 SEALs
Nuclear Weapons:	none
Conventional Weapons:	533mm (21in) torpedo
Sonar:	IBM BQQ 6
Navigation:	BPS 15A
Powerplant:	nuclear reactor, 15,000 shp
Date Commissioned:	1965 (relaunched 1993)

USS L. MENDEL RIVERS

The USS *L. Mendel Rivers* is a modified Sturgeon class SSN submarines that carries a Dry Dock Shelter (DDS) aft of its conning tower. The DDS is used to launch SEAL teams from the safety of submersion into enemy waters without detection. This method of SEAL delivery is remarkably effective and has been used in many conflicts around the world. The SEAL team's mission is often some form of intelligence gathering or covert infiltration prior to a more conventional attack. The United States Navy SEALs are amongst the best-trained and equipped special forces units in the world. Aboard a submarine, they live in cramped conditions preparing themselves for the mission ahead. They can deploy themselves in a manner of ways. They can use scuba equipment to swim to shore, an inflatable dingy or even a small submarine. Once ashore, the submarine moves to a safer location than the drop-off point and loiters, monitoring communications and relaying information. When the SEAL team has completed its mission, the submarine returns stealthily to a pre-arranged area where it can once more covertly collect the troops into the DDS. If everything goes to plan, the enemy would never be aware that its defences had been penetrated and that a US submarine and special operations team had been involved.

SPECIFICATIONS

Builder:	Newport News
Class:	SSN-637 Sturgeon
Number:	SSN-686
Mission:	attack/SEAL delivery submarine
Length:	92m (302ft)
Beam:	9.7m (32ft)
Displacement:	5039 tonnes (5119 tons)
Speed:	25 knots
Operating Depth:	350m (1200ft)
Maximum Depth:	600m (1980ft)
Crew:	107 + SEAL teams
Nuclear Weapons:	none
Conventional Weapons:	533mm (21in) torpedo, Harpoon
Sonar:	BQS-13 active sonar
Navigation:	BPS-14/15 radar
Powerplant:	nuclear reactor, 15,000 shp
Date Commissioned:	1975

USS LOS ANGELES

The Los Angeles class SSN was designed almost exclusively for carrier battlegroup escort; they were fast, quiet, and could launch Mk48 and ADCAP torpedoes, Harpoon Anti-Ship Missiles, and land-attack Tomahawk cruise missiles. Cutting edge when they were launched, the new submarines were another major improvement in noise-reduction technology and speed. Escort duties include conducting anti-submarine warfare sweeping hundreds of miles ahead of the carrier battlegroup and conducting attacks against enemy vessels. Submarines of the Los Angeles class are still among the most advanced undersea vessels of their type in the world. While anti-submarine warfare is still their primary mission, and given their original remit as a Cold War weapon, the inherent characteristics of the submarine's stealth, mobility and endurance are still more than qualified to meet the challenges of the twenty-first century's turbulent global geopolitical climate. The Los Angeles class are able to get on station quickly, stay for an extended period of time and carry out a variety of missions including the deployment of special forces, minelaying and precision strike land attack, all of which are important concepts of military operation in the new millenium. These boats are well equipped to accomplish these tasks, even though the design is from a past era.

SPECIFICATIONS

Builder:	Newport News
Class:	SSN-688 Los Angeles
Number:	SSN-688
Mission:	attack submarine
Length:	109.7m (360ft)
Beam:	10m (33ft)
Displacement:	6210 tonnes (6845 tons)
Speed:	32 knots
Operating Depth:	300m (990ft)
Maximum Depth:	400m (1320ft)
Crew:	129
Nuclear Weapons:	none
Conventional Weapons:	533mm (21in) torpedo, TLAM
Sonar:	BSY-1 sonar suite
Navigation:	AN/BPS-15 navigation radar
Powerplant:	nuclear reactor, 35,000 shp
Date Commissioned:	1976

USS LOUISIANA

The last Ohio class SSBN to be rolled out by the United States Navy is the USS *Louisiana*. In light of the changing nature of the global political situation and the end of the Cold War, the United States Government decided that the SSBN programme should be limited to 18 boats. SSBNs form a key part of the TRIAD strategic deterrence for the US, with land-based and air-launched nuclear weapons. The Ohio class of SSBNs are equipped with two different types of missiles, depending largely on the age of the boat. The USS *Louisiana* is fitted with one of the most sophisticated nuclear missiles systems in the world, the D-5 Trident II. The D-5 is essentially an evolution of its predecessor the C-4 Trident I, as carried by the older Ohio class submarines. Carrying its compliment of MIRVs in its nose cone, the D-5 can deliver them onto different targets up to 7360km (4600 miles) away. Able to carry 24 missiles, each with six to eight 500-kiloton MIRVs, one Ohio class SSBN in the Atlantic could destroy every European capital city, every state capital in the United States and still have warheads to spare. This awesome quantity of firepower is second to none in the modern world. There is more to the Ohio class than pure firepower, however, as it is also equipped with some of the most sensitive and effective sensors and communications equipment in the world.

SPECIFICATIONS

Builder:	*Electric Boat*
Class:	*SSBN-726 Ohio*
Number:	*SSBN-743*
Mission:	*ballistic missile submarine*
Length:	*170.7m (560ft)*
Beam:	*12.7m (42ft)*
Displacement:	*19,000 tonnes (18,750 tons)*
Speed:	*25 knots*
Operating Depth:	*300m (990ft)*
Maximum Depth:	*450m (1485ft)*
Crew:	*155*
Nuclear Weapons:	*24 x D-5 Trident II*
Conventional Weapons:	*533mm (21in) torpedo*
Sonar:	*BQS-13 active sonar*
Navigation:	*Raytheon navigation sonar*
Powerplant:	*nuclear reactor, 60,000 shp*
Date Commissioned:	*1997*

USS OHIO

The USS *Ohio* is the first ship in the Ohio class of fleet ballistic missile (FBM) submarines. This class replaced the ageing FBM submarines from the 1960s, and was brought into service during the early 1980s. The Ohio class submarine is vastly more capable than its predecessors, and is one of the most advanced pieces of technology on the planet. SSBN-726 class FBM submarines can carry 24 ballistic missiles each with 7 MIRV warheads that can be accurately delivered to selected targets from almost anywhere in the world's oceans. The hull has been designed to allow for greater speed whilst submerged, but also retaining an unprecedented level of stealth, making detection much harder. The larger hull of the Ohio class can accommodate more weapons of larger size and greater range, as well as sophisticated computerized electronic equipment for improved weapon guidance and sonar performance. The Ohio class submarines are specifically designed to be able to operate at sea for exceptionally long periods of time. To reduce the time needed in port for replenishment, and also to make such processes easier, three large logistics hatches are fitted to provide large-diameter resupply and repair openings. These hatches allow the rapid transfer of supply pallets, equipment replacement modules and machinery.

SPECIFICATIONS

Builder:	Electric Boat
Class:	SSBN-726 Ohio
Number:	SSBN-726
Mission:	ballistic missile submarine
Length:	170.7m (560ft)
Beam:	12.7m (42ft)
Displacement:	19,000 tonnes (18,750 tons)
Speed:	25 knots
Operating Depth:	300m (990ft)
Maximum Depth:	450m (1485ft)
Crew:	155
Nuclear Weapons:	24 x C-4 Trident I
Conventional Weapons:	533mm (21in) torpedo
Sonar:	BQQ-6 passive sonar
Navigation:	Raytheon navigation sonar
Powerplant:	nuclear reactor, 60,000 shp
Date Commissioned:	1981

USS PARCHE

Sturgeon class submarines were built for anti-submarine warfare in the late 1960s and 1970s. Using the same propulsion system as the predecessor SSN-585 Skipjack and SSN-594 Permit classes, the larger Sturgeons sacrificed speed for greater combat capabilities. They have been modified to carry the Harpoon missile, the Tomahawk cruise missile, as well as the Mk48 and ADCAP torpedoes. The torpedo tubes are located amidships to accommodate the bow-mounted sonar. The sail-mounted dive planes rotate to a vertical position for breaking through the ice when surfacing in Arctic regions. Beginning with SSN-678 Archerfish, units of this class had a longer hull, giving them more living and working space than previous submarines of the Sturgeon Class. A total of six Sturgeon class boats have been modified to carry the SEAL Dry Deck Shelter (DDS), one in 1982 and five between 1988 and 1991. In this configuration they are primarily tasked with the covert insertion of special forces troops from the attached DDS. The Dry Deck Shelter is a submersible launch hanger with a hyperbaric pressure chamber that attaches to the ship's weapon shipping hatch. Rapidly being phased out in favour of the Los Angeles and Seawolf class vessels, this venerable and flexible workhorse of the submarine attack fleet continues to operate to this day.

SPECIFICATIONS

Builder:	Ingalls Shipbuilding
Class:	SSN-637 Sturgeon
Number:	SSN-683
Mission:	attack/SEAL delivery submarine
Length:	92m (302ft)
Beam:	9.7m (32ft)
Displacement:	5039 tonnes (5119 tons)
Speed:	25 knots
Operating Depth:	300m (990ft)
Maximum Depth:	350m (1155ft)
Crew:	107
Nuclear Weapons:	none
Conventional Weapons:	533mm (21in) torpedo, TLAM
Sonar:	BQS-13 active sonar
Navigation:	BPS-14/15 radar
Powerplant:	nuclear reactor, 15,000 shp
Date Commissioned:	1974

USS SANTA FE

The USS *Santa Fe* is an improved version of the Los Angeles class submarine, and one of the most modern Los Angeles class boats in US Navy service. All the improved versions of the Los Angeles class SSN are equipped with the Tomahawk land attack (TLAM) cruise missile. The addition of this weapons system to the SSN is a vital step in the evolution of the attack submarine. Since the end of the Cold War, the Los Angeles class submarine has lost its primary enemy as the Russian threat has disappeared and its ballistic and attack submarines have all but fallen into disrepair. Finding another role for the billions of dollars of SSN capability has been one of the prime focuses in recent years. The addition of the Tomahawk cruise missile appears to have solved the problem. Whilst continuing to fulfil its role as protection for a carrier group, it can also take part in force projection and precision strike missions with the Tomahawk cruise missile. This modern submarine can sneak undetected silently into enemy waters, and launch a Tomahawk whilst still submerged at a target over 1200km (750 miles) into enemy territory. This ability to strike at a target with great accuracy whilst remaining hidden is a huge advantage in modern warfare. It is also a great assistance in the peace-enforcement or peacekeeping operations that the United States takes part in.

SPECIFICATIONS

Builder:	Electric Boat
Class:	SSN-688 Improved Los Angeles
Number:	SSN-763
Mission:	attack submarine
Length:	109.7m (360ft)
Beam:	10m (33ft)
Displacement:	6210 tonnes (6845 tons)
Speed:	32 knots
Operating Depth:	300m (990ft)
Maximum Depth:	400m (1320ft)
Crew:	129
Nuclear Weapons:	none
Conventional Weapons:	533mm (21in) torpedo, TLAM
Sonar:	BSY-1 sonar suite
Navigation:	AN/BPS-15 navigation radar
Powerplant:	nuclear reactor, 35,000 shp
Date Commissioned:	1993

USS SEAWOLF

The USS *Seawolf* is the first of a new class of submarines designed to operate autonomously against the world's most capable submarine and surface threats. The primary mission of the Seawolf would have been to destroy Soviet ballistic missile submarines before they could attack American targets, had the Cold War not ended in 1991. The Soviet submarines were and remain one of the most effective elements of their intercontinental ballistic missile arsenal, but they no longer present a genuine threat to US interests. In addition to their capabilities in countering enemy submarines and surface shipping, Seawolf submarines are suited for battlespace-preparation roles, and are thus useful in the post-Cold War era. Incorporation of sophisticated electronics gives the Seawolf enhanced indications and warning, surveillance and communications capabilities that are extremely useful in a battlegroup. These platforms are capable of integrating into a battlegroup's infrastructure, providing support and protection, or shifting rapidly into a land battle support role with its Tomahawk land-attack missiles (TLAM). Seawolf also incorporates the latest in quiet technology to keep pace with the threat then posed by an aggressive Soviet Union. It is said that the Seawolf class is quieter at its tactical speed of 25 knots than a Los Angeles class submarine at pierside.

SPECIFICATIONS

Builder:	Electric Boat
Class:	SSN-21 Seawolf
Number:	SSN-21
Mission:	attack submarine
Length:	107.6m (353 ft)
Beam:	12.2m (40 feet)
Displacement:	9137 tonnes (9283 tons)
Speed:	35 knots
Operating Depth:	400m (1325ft)
Maximum Depth:	classified
Crew:	133
Nuclear Weapons:	none
Conventional Weapons:	660mm (30in) torpedo, TLAM
Sonar:	BSY-2 sonar suite
Navigation:	BPS-16 navigation radar
Powerplant:	nuclear reactor, 52,000 shp
Date Commissioned:	1997

USS TENNESSEE

The USS *Tennessee* SSBN is an Ohio class submarine fitted with the D-5 Trident II nuclear weapon. As part of the United States TRIAD system of nuclear deterrence, the Ohio class submarine must be able to deliver its firepower onto its target within minutes of receiving a launch instruction. To this end the USS *Tennessee* must be within range of its target. Designed and built during the unstable days of the Cold War, the Ohio class submarines were designed to patrol the seas within range of the Soviet Union, able to launch at a moment's notice. All the while Soviet hunter submarines would be searching for them. The Soviet Akula class attack submarines were designed specifically to counter the threat of US ballistic missile submarines, and the two vessels played a game of cat-and-mouse under the Polar ice caps for much of the 1980s. Thus the Ohio was designed to be especially quiet. Even though it is some years old and technology has moved forward, the Ohio class is still one of the quietest submarines in the world. Much of the technology and design that makes it so stealthy is top secret to this day, and whilst the original enemy has disappeared, the Ohio class is still able to patrol hostile waters without detection, collecting intelligence and keeping America's interests abroad safe from hostile intent. USS *Tennessee* is expected to serve until 2020.

SPECIFICATIONS

Builder:	Electric Boat
Class:	SSBN-726 Ohio
Number:	SSBN-734
Mission:	ballistic missile submarine
Length:	170.7m (560ft)
Beam:	12.7m (42ft)
Displacement:	19,000 tonnes (19,304 tons)
Speed:	25 knots
Operating Depth:	300m (990ft)
Maximum Depth:	450m (1485ft)
Crew:	155
Nuclear Weapons:	24 x D-5 Trident II
Conventional Weapons:	533mm (21in) torpedo
Sonar:	BQS-13 active sonar
Navigation:	Raytheon navigation sonar
Powerplant:	nuclear reactor, 60,000 shp
Date Commissioned:	1988

USS TEXAS

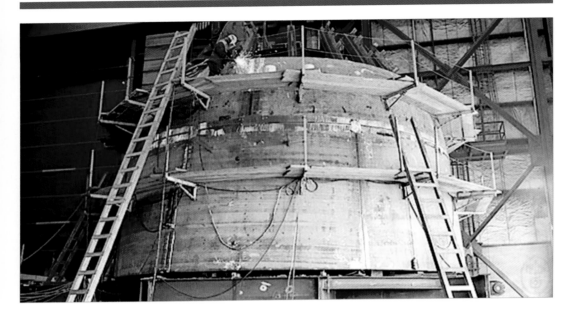

The Virginia class submarines, of which the USS *Texas* will be one of the earliest examples, is arguably a quantum leap in submarine technology and doctrine. Taking much of its design and development directly from the halted Seawolf class submarine, the Virginia class will stand head and shoulders above any of its rivals across the globe. The most innovative feature of the Virginia class is undoubtedly the integral lock-in/lock-out chamber. This will be used for delivering special forces teams from the boat in much the same way as the Dry Dock Shelter (DDS) systems do. However, the difference is that the Virginia class submarines will host the new Advanced SEAL Delivery System (ASDS), a miniature submarine for covert use by special forces. The chamber will also give the Virginia class much greater flexibility and size in payload than the DDS system does. This evolution of the DDS system into an attack submarine highlights the US Navy's desire for greater operational flexibility and capability from its submarine force. Along with the traditional attack role and the new special warfare role, the Virginia class will also be able to attack land targets with their Tomahawk cruise missiles. The integration of all these different systems into one submarine marks a shift in US operational doctrine, and undoubtedly leads the way for future submarine development.

SPECIFICATIONS

Builder:	Newport News
Class:	SSN-774 Virginia
Number:	SSN-775
Mission:	attack submarine
Length:	114m (377ft)
Beam:	10.3m (34ft)
Displacement:	7800 tonnes (7924 tons)
Speed:	28 knots
Operating Depth:	400m (1320ft)
Maximum Depth:	classified
Crew:	113
Nuclear Weapons:	none
Conventional Weapons:	533mm (21in) torpedo, TLAM
Sonar:	TB-29 towed array
Navigation:	BPS-16 navigation radar
Powerplant:	S9G nuclear reactor
Date Commissioned:	2005

USS VIRGINIA

The USS *Virginia* is the first in the new class of attack submarine. It is an advanced stealth multi-mission nuclear-powered submarine for deep-ocean anti-submarine warfare and for littoral (shallow water) operations. Although the Seawolf submarine was developed to provide an eventual replacement for the US Navy Los Angeles class submarines in combating Soviet forces, the prohibitive unit cost and changing strategic requirements led to the US Navy defining a smaller new-generation attack submarine. The noise level of the Virginia is equal to that of the US Navy Seawolf, SSN 21, with a lower acoustic signature than the Russian Improved Akula class and the proposed Russian Fourth Generation Attack Submarines. To achieve this low acoustic signature, the Virginia incorporates newly designed anechoic coatings, isolated deck structures and a new design of propulsor. The submarine is equipped with 12 vertical missile-launch tubes and four 533mm (21in) torpedo tubes. The vertical launching system has the capacity to launch 16 Tomahawk submarine launched cruise missiles in a single salvo. There is capacity for up to 26 Mk48 ADCAP Mod 6 torpedoes, and Sub Harpoon anti-ship missiles can be fired from the torpedo tubes. Mk60 CAPTOR mines may also be fitted. The USS *Virginia* also carries the Advanced SEAL Delivery System (ASDS).

SPECIFICATIONS

Builder:	Electric Boat
Class:	SSN-774 Virginia
Number:	SSN-774
Mission:	attack submarine
Length:	114m (377ft)
Beam:	10.3m (34ft)
Displacement:	7800 tonnes (7924 tons)
Speed:	28 knots
Operating Depth:	400m (1325ft)
Maximum Depth:	classified
Crew:	113
Nuclear Weapons:	none
Conventional Weapons:	533mm (21in) torpedo, TLAM
Sonar:	TB-29 Towed Array
Navigation:	BPS-16 navigation radar
Powerplant:	S9G nuclear reactor
Date Commissioned:	2004

HMAS ANZAC

This Anzac class frigate is a joint Australia–New Zealand project for a total of 10 warships. The design is based on the Blohm & Voss Meko 200 modular design which utilizes a basic hull and construction concept to provide flexibility in the choice of command and control, weapons, equipment and sensors. The *Anzac* is armed with one eight-cell Mk 41 Vertical Launching System (VLS) for NATO Sea Sparrow surface-to-air missiles (SAMs). Sea Sparrow is a semi-active radar missile with a range of 14.5km (9 miles). The capacity to launch eight Boeing Harpoon anti-ship missiles (ASMs) is also to be added. The main gun is a United Defense 127mm Mk 45 Model 2, which can fire at a rate of 20 rounds a minute up to a range of 20km (12.43 miles). Two triple 324mm Mk 32 torpedo tubes for Mk 46 anti-submarine torpedoes are also fitted to the ship. The Mk 46 is an active/passive torpedo with a range of 11km (6.83 miles). This state-of-the-art frigate is equipped with a Sceptre A radar warner and will be fitted with a radar jammer which as yet does not have an official name. Decoy systems on the ship consist of Mk 36 launchers for Sea Gnat decoys and SLQ-25A towed torpedo decoys. Each ship is designed to accommodate, operate and maintain a Sikorsky S-70B2 helicopter, to be replaced later by the Kaman SH-2G Super Seasprite.

SPECIFICATIONS

Type:	frigate
Crew:	163
Displacement:	3658 tonnes (3600 tons)
Length:	118m (387.13ft)
Beam:	14.8m (48.55ft)
Draught:	4.35m (14.27ft)
Speed:	27 knots
Range:	9654km (6000 miles)
Missiles:	Sea Sparrow SAM, Harpoon ASM
Guns:	1 x 127mm
Torpedoes:	Mk 46
Helicopters:	1 x S-70B2
Aircraft:	none
Air Search Radar:	SPS-49 (V) 8
Surface Radar:	CelsiusTech 9LV 453 TIR
Fire Control Radar:	unknown
Propulsion:	2 x diesels, 30,000shp

COLOSSUS

An ex-Royal Navy light fleet carrier acquired in 1956, the *Colossus* is the Brazilian Navy's largest ship, and despite its considerable age, remains in service. She will be replaced by the *Sao Paulo* (the former French carrier acquired in 2000) and decommissioned by 2003. Originally built by Swan Hunter, it was laid down on 16 November 1942, launched on 23 February 1944 and commissioned on 15 January 1945. She served in the Mediterranean, then in the Pacific, participated in cold-weather trials in the Arctic in 1948–49, and served as a troop and aircraft transport in 1951–52. Her designation changed to R71 under the NATO designation system. Refitted in the early 1950s and loaned to Australia on 13 November 1952 as HMAS *Vengeance*, she was sold to Brazil on 12 December 1956. Refitted at Rotterdam between 1957 and 1960, she was recommissioned on 6 December 1960 as the *Minas Gerais* (A11). Refitted in 1976–80, she was laid up in 1987 due to catapult problems. She was refitted again between 1991 and 1993, her catapult becoming operational in 1996. During the 1990s she operated only anti-submarine warfare (ASW) helicopters, but A-4 Skyhawk aircraft have since been acquired. Her armament includes Mistral surface-to-air missiles (SAMs) and 40mm and 47mm guns.

SPECIFICATIONS

Type:	light aircraft carrier
Crew:	1300
Displacement:	20,208 tonnes (19,890 tons)
Length:	211.8m (694.88ft)
Beam:	24.4m (80ft)
Draught:	7.5m (24.6ft)
Speed:	24 knots
Range:	19,308km (12,000 miles)
Missiles:	Mistral SAM
Guns:	10 x 40mm, 2 x 47mm
Torpedoes:	none
Helicopters:	6 x H-3, 2 x UH-1, 3 x Super Puma
Aircraft:	6 x S-2G Trackers
Air Search Radar:	SPS 40B
Surface Radar:	Plessey AWS 4
Fire Control Radar:	2 x SPG 34
Propulsion:	steam turbines, 40,000shp

HALIFAX

Incorporating many technological advances, including an integrated communications system, a command and control system, and a machinery control system, the Halifax class multirole frigates' weapons, sensors and engines form a significant platform of defensive and offensive capabilities. These ships are quiet, fast and have excellent sea-keeping characteristics. Halifax class frigates deploy singly or as part of a task group anywhere in the world with NATO ships, US carrier battle groups or in cooperation with other allied vessels. These deployments would not be possible without a fleet of modern, versatile patrol frigates. The frigates carry a formidable array of weapons and sensor systems including eight Harpoon long-range surface-to-surface missiles (SSMs) and also termed anti-ship missiles (ASMs), 16 Sea Sparrow surface-to-air missiles (SAMs), a Bofors 57mm rapid-fire gun, a 20mm Phalanx anti-missile close-in weapons system (CIWS), eight 12.7mm machine guns and 24 anti-submarine homing torpedoes. In addition, the ships can defend themselves using infrared suppression, Shield decoys, chaff, flares, a towed acoustic decoy, and radar and sonar jamming devices. Finally, each ship's torpedo-carrying Sea King helicopter significantly extends its overall range of operational effectiveness.

SPECIFICATIONS

Type:	multirole frigate
Crew:	225
Displacement:	5319 tonnes (5235 tons)
Length:	134.7m (441.92ft)
Beam:	16.4m (53.8ft)
Draught:	4.9m (16ft)
Speed:	28 knots
Range:	11,424km (7100 miles)
Missiles:	Harpoon ASM, Sea Sparrow SAM
Guns:	1 x 57mm, 1 x Phalanx
Torpedoes:	Mk 32
Helicopters:	1 x Sea King
Aircraft:	none
Air Search Radar:	SPS 49 (V) 5
Surface Radar:	Ericsson Sea Giraffe HC 150
Fire Control Radar:	2 x Signaal VM 25 STIR
Propulsion:	1 x diesel, 2 x turbines, 47,494shp

IROQUOIS

After years as a pre-eminent anti-submarine warfare vessel during the latter stages of the Cold War, *Iroquois* and her three sisters – *Huron*, *Athabaskan* and *Algonquin* – were refitted in the early 1990s as "command and control ships" with upgraded anti-aircraft defences and improved communications and sensor systems. *Iroquois* and *Algonquin* were further upgraded in the late 1990s. In the aftermath of the 1982 Falklands War (where British warships were sunk by bombs dropped from aircraft and sea-skimming missiles) and the war in the Persian Gulf in 1991 where the Allied coalition fleet came under attack from Chinese-made Silkworm surface-to-surface missiles (SSMs), addressing the problem of air defence became a top priority for Western navies. The Canadian Navy decided to convert the four 280s to the area air defence role. This gave them the self-defensive capabilities they needed to become command and control ships. Thus equipped, the navy could form independent task groups, responsible for their own security. As well as substantial changes to the superstructure of the vessels, new air defence weapons were installed including standard vertically launched missiles and a 76mm Super Rapid gun. The Phalanx close-in weapons system (CIWS) provides a final defence against sea-skimming missiles.

SPECIFICATIONS

Type:	air defence destroyer
Crew:	280
Displacement:	5020 tonnes (5100 tonnes)
Length:	129.8m (425.85ft)
Beam:	15.2m (49.86ft)
Draught:	4.7m (15.41ft)
Speed:	29 knots
Range:	7240km (4500 miles)
Missiles:	Mk 41 SAM
Guns:	1 x 76mm, 1 x 20mm, 1 x Phalanx
Torpedoes:	Mk 46
Helicopters:	2 x Sea King
Aircraft:	none
Air Search Radar:	Signaal LW 08
Surface Radar:	Signaal DA 08
Fire Control Radar:	Signaal STIR 1 8
Propulsion:	2 x gas turbines, 50,000shp

LUDA

The Chinese Type 051 Luda class guided missile destroyers are similar to the discontinued Soviet Kotlin class. Intended for anti-ship missions, their primary offensive armament is a pair of "Sea Eagle I" ship-to-ship missiles (SSMs). However, the combat potential provided by the high speed and long range of this ship design was compromised by the lack of an air defence ability. In common with the old Soviet Kotlin class, the Luda class ships originally had no surface-to-air missiles (SAMs) for self-protection, a deficiency which the Soviet Navy eventually remedied with the SAM Kotlin class. Three major Luda variants have been produced: Luda I had an initial basic configuration, though with considerable variations in armament and electronics among its various units; the Luda II added the eight-cell HQ-7 SAM system, along with a helicopter deck and hangar replacing the aft gun armament; and the Luda III features improved sonars, SSMs and electronics on a single ship, the DD166 *Zhuhai*. The initial Luda I design has now been extensively modernized, primarily for anti-submarine warfare missions, becoming Luda III class. However, thus far only one vessel has been identified (hull number 166, possibly changed to 168 for overseas deployment), but more Luda Is are expected to be upgraded to the Luda III standard in the future.

SPECIFICATIONS

Type:	*guided missile destroyer*
Crew:	*280*
Displacement:	*3729 tonnes (3670 tons)*
Length:	*132m (433ft)*
Beam:	*12.8m (41.99ft)*
Draught:	*4.6m (15.09ft)*
Speed:	*32 knots*
Range:	*4779km (2970 miles)*
Missiles:	*Sea Eagle SSM, Crotale SAM*
Guns:	*4 x 130mm, 8 x 57, 37 & 25mm*
Torpedoes:	*Whitehead A 244S*
Helicopters:	*2 x Harbin Z-9A*
Aircraft:	*none*
Air Search Radar:	*Knife Rest or Cross Slot*
Surface Radar:	*Eye Shield or Sea Tiger*
Fire Control Radar:	*Wasp Head or Type 343 Sun Visor B*
Propulsion:	*steam turbines, 60,000shp*

CASSARD

The French Navy's Cassard class destroyer is equipped with the Senit combat data system and an OPSMER command support system. Senit gathers, correlates, evaluates and displays information from shipboard sensors and handles data exchanges with other units, via tactical data links, including Link 11 and 14 and the Syracuse satellite communications system. Two four-cell missile launchers for the Exocet MM40 anti-ship missile (ASM) are installed in a midship position between the two citadels. Two Sadral six-round turrets for the Mistral surface-to-air missile (SAM) are mounted on the raised deck each side of the helicopter hangar. The infrared-guided Mistral provides short-range air defence up to a range of 5km (3.1 miles). The main gun is the DCN 100mm model, which is capable of firing at a rate of 80 rounds per minute up to a range of 8km (5 miles). The ship also has two Oerlikon 20mm guns, which have a range of 10km (6.25 miles) and a firing rate of 720 rounds per minute. The ship has a flight deck at the stern with a single landing spot for the AS 565MA Panther helicopter. To facilitate all-weather operations, the DCN Samahe helicopter handling system allows deployment and recovery of the aircraft in rough seas. In addition, the ship stores an arsenal of helicopter-launched Mark 46 torpedoes.

SPECIFICATIONS

Type:	destroyer
Crew:	245
Displacement:	4775 tonnes (4700 tons)
Length:	139m (456ft)
Beam:	14m (45.93ft)
Draught:	6.5m (21.32ft)
Speed:	29.5 knots
Range:	13,194km (8200 miles)
Missiles:	Exocet ASM, Mistral SAM
Guns:	1 x 100mm, 2 x 20mm
Torpedoes:	KD 59E, ECAN L5 Mod 4
Helicopters:	1 x AS 565MA Panther
Aircraft:	none
Air Search Radar:	DRBJ 11 B, 3D
Surface Radar:	DRBV 26C
Fire Control Radar:	DRBC 33A
Propulsion:	4 x diesels, 43,200shp

CASSIOPÉE

French mine warfare units are tasked with permanently securing the approaches to the Brest Ship, Submersible, Ballistic, Nuclear (SSBN) base, one of France's strategic ports. In the event of particular threats they must also be in a position to simultaneously keep open the ports on the Channel–Atlantic seaboard, in Toulon and Marseilles-Fos, and to keep forces on standby to ensure free access to any and all allied ports. They also constitute a major component for operations involving projection of forces inshore, which are more sensitive to the mine threat. In order to fulfil these missions, the Naval Action Force includes minehunters fitted with the equipment required to identify and destroy devices laid on the sea bed, teams of clearance divers who operate in shallow waters, and sonar tugs to guard the approaches to Brest itself. The *Cassiopée* is one of the Eridan class of minehunters, which have a minimal light minesweeping capability. In January 2000, Directions des Constructions Navales and Thomson Marconi Sonar SAS were awarded parallel contracts to update all 13 of the French Navy's Eridan class "Tripartite" minehunters with a propelled variable depth sonar (PVDS) capability, a new hull-mounted minehunting sonar and a new mine warfare tactical system.

SPECIFICATIONS

Type:	minehunter
Crew:	49
Displacement:	605 tonnes (595 tons)
Length:	51.6m (169.29ft)
Beam:	9m (29.52ft)
Draught:	3.5m (11.48ft)
Speed:	15 knots
Range:	4827km (3000 miles)
Missiles:	none
Guns:	1 x 20mm, 2 x 12.7mm
Torpedoes:	none
Helicopters:	none
Aircraft:	none
Air Search Radar:	none
Surface Radar:	Decca 1229
Fire Control Radar:	none
Propulsion:	1 x diesel, 1900shp

CHARLES DE GAULLE

The nuclear-powered aircraft carrier *Charles De Gaulle* was constructed at the Brest Naval Shipyard in Brittany. The ship was launched in May 1994 and commissioned in September 2000, following sea trials which began in January 1999. The ship operates a fleet of 40 aircraft: Rafale M, Super Etendard and three E-2C Hawkeye airborne early warning (AEW) aircraft. The ship will also support the AS 565 Panther or NH 90 helicopter. The main deck consists of a main runway angled at 8.5 degrees to the ship's axis and an aircraft launch area forward of the island. These are each equipped with a US Navy Type C13 catapult, capable of launching one aircraft per minute. The carrier is fitted with the SATRAP computerized, integrated stabilization system designed to maintain stabilization to within 0.5 degrees of horizontal, allowing aircraft to operate in rough seas. The Aster 15 surface-to-air missile (SAM) provides defence against hostile aircraft and anti-ship missiles. Two eight-cell Sylver vertical launch systems are installed on the starboard side forward of the bridge and two on the port side aft of the bridge. The ship has two six-cell Sadral launching systems for the Mistral SAM. The nuclear propulsion system has the capacity to provide five years of continuous operation at 25 knots before refuelling.

SPECIFICATIONS

Type:	aircraft carrier
Crew:	1950 (including air wing)
Displacement:	38,000 tonnes (37,401 tons)
Length:	216.5m (710.3ft)
Beam:	31.5m (103.34ft)
Draught:	8.45m (27.72ft)
Speed:	25 knots
Range:	unlimited
Missiles:	Aster 15 SAM, Mistral SAM
Guns:	8 x 20mm
Torpedoes:	none
Helicopters:	see below
Aircraft:	40 (including helicopters)
Air Search Radar:	DRBV 15 3-D, DRBJ 11B
Surface Radar:	DRBV 15C Sea Tiger Mark 2
Fire Control Radar:	Arabel missile control
Propulsion:	nuclear, 83,000shp

FLOREAL

The *Floreal* is the lead ship of a class of French Navy low-cost patrol vessels for overseas service and fisheries patrol. Officially described as surveillance ships designed to operate offshore in low-intensity operations, they have been built to civilian commercial standards, yet appear to provide adequate military features such as survivability (it is important to remember that the manufacturing regulations for the construction of dangerous cargo ships and the quality required for passenger liners are very strict and exhaustive; indeed, they are very similar to those required of naval ships). The vessels of this class feature the modular construction of systems and subsystems throughout, and particularly in the main and auxiliary machinery spaces. Once put together the modules are well arranged and compact, yet items are accessible at all times. Passage around each module on the boat for crew members is easy and does not require manoeuvring around protruding objects. As well as the crew of 80 sailors and officers, the ship has room to accommodate up to 24 special forces personnel. The aft helicopter deck and hangar is for the use and storage of one Panther helicopter. There are six vessels in this class: F730 *Floreal*, F731 *Prairial*, F732 *Nivose*, F733 *Ventose*, F734 *Vendermaire* and F735 *Germinal* (shown above).

SPECIFICATIONS

Type:	patrol frigate
Crew:	80
Displacement:	2997 tonnes (2950 tons)
Length:	93.5m (306.75ft)
Beam:	14m (45.93ft)
Draught:	4.3m (14.1ft)
Speed:	20 knots
Range:	16,090km (10,000 miles)
Missiles:	Exocet SSM
Guns:	1 x 100mm, 2 x 20mm
Torpedoes:	none
Helicopters:	1 x Panther
Aircraft:	none
Air Search Radar:	DRBV 21A
Surface Radar:	DRBV 21A
Fire Control Radar:	none
Propulsion:	4 x diesels, 8800shp

FOCH

The carriers of the Clemenceau class have been France's premier naval ships since World War II. The *Clemenceau* was decommissioned in October 1997 and the *Foch* will either be held in reserve as a back-up for the new *Charles De Gaulle* nuclear carrier (see page 13) or transferred to the Brazilian Navy for further service there. The ship's air wing includes the Super Etendard, which is configured to carry the ASMP (*Air-Sol Moyenne Portée*) nuclear missile, which is powered by a ramjet with an integrated accelerator. Armed with a tactical nuclear warhead, the ASMP is produced by Aerospatiale, except for the military head, which is provided by the Atomic Energy Commission. The ASMP's nuclear warhead has five times the power of the freefall weapons it replaces. This supersonic missile is guided by a stand-alone system of inertial navigation that provides it with the precision required and allows the launcher aircraft to remain a safe distance from the enemy defences (which is a bonus for the slow-moving Super Etendard). The missile's propulsion system consists of a statoréacteur using liquid fuel developed by Aerospatiale. ASMP became operational in May 1986 on Mirage IVP aircraft and beginning in 1988 on Mirage 2000N aircraft; and it is also being adapted on Super Etendard aircraft for the French Navy.

SPECIFICATIONS

Type:	aircraft carrier
Crew:	1338 plus 582 air wing
Displacement:	33,304 tonnes (32,780 tons)
Length:	265m (869.42ft)
Beam:	31.7m (104ft)
Draught:	8.6m (28.21ft)
Speed:	32 knots
Range:	12,067km (7500 miles)
Missiles:	Crotale SAM
Guns:	4 x 100mm
Torpedoes:	none
Helicopters:	4 (40 for amphibious assault)
Aircraft:	35
Air Search Radar:	DRBV 23B
Surface Radar:	DRBI 10, 1 DRBV 15
Fire Control Radar:	2 x DRBC 32B
Propulsion:	steam turbines, 126,000shp

FOUDRE

The French Navy operates two Foudre class landing platform dock ships, the *Foudre* (L9011) and the *Siroco* (L9012), which were commissioned in 1990 and 1998 respectively. They were built at the Brest Naval Dockyard. The ships are assigned to the Naval Action Force based at the Toulon Mediterranean Command. The Foudre class is capable of landing and supporting a mechanized armoured regiment of the French rapid deployment force. The three main missions of the Foudre class are the landing of infantry and armoured vehicles on unprepared coasts, mobile logistic support for naval forces, and humanitarian missions. The ship's complement is 210 crew with 13 officers; it can also accommodate 467 passengers or troops. With 700 crew and passengers, the ship has an endurance of 30 days. In times of crisis the ship can accommodate up to 1600 people. To meet military and humanitarian requirements, Foudre class ships provide hospital facilities for large-scale medical and evacuation missions, including two fully equipped operating theatres and 47 beds. The ship can accommodate up to seven Super Puma helicopters and the flight deck is equipped with a Samahe haul-down system. The helicopter hangar has capacity for two Super Frelon helicopters or up to four Super Puma helicopters.

SPECIFICATIONS

Type:	landing platform dock
Crew:	223
Displacement:	12,090 tonnes (11,900 tons)
Length:	168m (551.18ft)
Beam:	23.5m (77ft)
Draught:	5.2m (17ft)
Speed:	21 knots
Range:	17,699km (11,000 miles)
Missiles:	Simbad SAM
Guns:	1 x 40mm, 2 x 20mm
Torpedoes:	none
Helicopters:	7 x Super Puma
Aircraft:	none
Air Search Radar:	DRBV 21A
Surface Radar:	Racal Decca 2459
Fire Control Radar:	none
Propulsion:	2 x diesels, 21,600shp

GEORGES LEYGUES

The Georges Leygues class frigates are assigned the general roles of protection and defence of French and European maritime interests in coordination with allies; ensuring the military presence of France all over the world and the protection of French citizens worldwide; public service missions such as surveillance and rescuing of people and property; and general humanitarian operations. Designed mainly to maintain and enforce French interests in overseas maritime areas and to participate in crisis management outside Europe, these warships are also integrated into the French Carrier Group formation when required. They may thus be used, as part of these missions, to provide support for an intervention force, protection for commercial traffic, or to carry out special operations. The importance attached to reducing their radar and acoustic signatures during construction, their modular design and their high degree of automation make them true twenty-first century ships, innovative in more ways than one. At sea their equipment and weapons are more specifically suited to fighting conventional submarines, especially in shallow waters. The increasing number of submarines deployed by secondary navies throughout the world, plus the increase in amphibious operations, make these warships highly important to the French Navy.

SPECIFICATIONS

Type:	frigate
Crew:	235
Displacement:	4420 tonnes (4350 tons)
Length:	139m (456ft)
Beam:	14m (45.93ft)
Draught:	5.7m (18.7ft)
Speed:	30 knots
Range:	13,676km (8500 miles)
Missiles:	Exocet SSM, Crotale SAM
Guns:	1 x 100mm, 2 x 20mm
Torpedoes:	Mk 46
Helicopters:	2 x Sea Lynx
Aircraft:	none
Air Search Radar:	DRBV 26
Surface Radar:	DRBV 51C
Fire Control Radar:	Thomson-CSF Vega with DRBC 32E
Propulsion:	2 x diesels, 52,000shp

JEANNE D'ARC

A combination of helicopter ship and cruiser, *Jeanne d'Arc* is a training ship in peacetime and an anti-submarine warfare (ASW) or assault ship in wartime. Originally intended only for training to replace a prewar cruiser, she was redesigned with ASW and assault capabilities due to a general shuffling of duties among the French carriers. *Jeanne d'Arc* has a large helicopter deck aft with a hangar below, and a midship superstructure with cruiser-type weapons mounted forward. She is lightly armed, reflecting her training role. During construction the gun armament was reduced, and an ASW rocket launcher eliminated. The hangar is used for berthing and other functions in peacetime but can be quickly reconverted for aircraft if needed. In 1975 six Exocet surface-to-surface missiles (SSMs) were fitted. Exocet missiles started development in 1967, originally as the ship-launched variant MM 38 which entered service in 1975. The air-launched version, AM 39, was developed later starting in 1974 and entering service with the French Navy in 1979. The missile is designed to attack large warships. A block 2 upgrade programme was carried out from the late 1980s until 1993, which introduced an improved digital active radar seeker and upgraded inertial navigation and control electronics.

SPECIFICATIONS

Type:	helicopter carrier
Crew:	696 plus 140 cadets
Displacement:	13,482 tonnes (13,270 tons)
Length:	182m (597.11ft)
Beam:	24m (78.74ft)
Draught:	7.3m (23.95ft)
Speed:	26 knots
Range:	9654km (6000 miles)
Missiles:	Exocet SSM
Guns:	4 x 100mm
Torpedoes:	none
Helicopters:	4 x Alouette, 8 x Super Puma
Aircraft:	none
Air Search Radar:	DRBV 22D
Surface Radar:	DRBV 51
Fire Control Radar:	3 x DRBC 32A
Propulsion:	1 x steam turbine, 40,000shp

LA MOTTE-PICQUET

This Georges Leygues class frigate is named after the French Admiral Toussaint Guillaume Picquet De La Motte (1720–91), who fought the English in the Caribbean during the American Revolutionary War. He was victorious in several key engagements and made an important visit to Savannah, Georgia, in 1779 in support of American independence. FS *La Motte-Picquet* (D645), commissioned in 1988, is a frigate specializing in anti-submarine warfare. The ship carries up to 26 Crotale surface-to-air missiles (SAMs). The Crotale system provides air situation and threat assessment, extended detection range, Identification Friend-or-Foe (IFF), multi-target detection plus automated acquisition, tracking and engagement and all-weather operation. The Crotale NG VT1 missile features a high level of manoeuvrability: speed is Mach 3.5, using a solid propellant rocket motor. The VT1 has an effective range of about 11km (6.87 miles) and ceiling of over 6000m (19,685ft). The command to line-of-sight (CLOS) guidance uses radar and electro-optical sensors. The missile is armed with a focused blast and fragmentation warhead, initiated by a proximity fuse. The warhead provides a lethal blast radius of 8m (26.24ft). Typical interception time from firing to airborne target destruction at a distance of 8km (5 miles) is 10.3 seconds.

SPECIFICATIONS

Type:	frigate
Crew:	235
Displacement:	4420 tonnes (4350 tons)
Length:	139m (456ft)
Beam:	14m (45.93ft)
Draught:	5.7m (18.7ft)
Speed:	30 knots
Range:	13,676km (8500 miles)
Missiles:	Exocet SSM, Crotale SAM
Guns:	1 x 100mm, 2 x 20mm
Torpedoes:	Mk 46
Helicopters:	2 x Lynx
Aircraft:	none
Air Search Radar:	DRBV 26
Surface Radar:	DRBV 51C
Fire Control Radar:	DRBC 32E
Propulsion:	2 x diesels, 52,000shp

L'AUDACIEUSE

The L'audacieuse class of offshore patrol boats is designed for overseas service. The main armament is a 40mm gun – an all-round weapon system with high development potential, and a gun that can be used with equal lethality in the traditional roles of field air defence or for combating ground threats. The gun fires a Bofors 3P round, which is state-of-the-art in terms of proximity fuzed ammunition: a multi-role round so versatile that it can be used against all targets within the aerial threat concept – aircraft, helicopters and stand-off weapons – as well as ground targets with the same lethal effect. The 3P round has an increased tungsten pellet payload and is the heaviest, most powerful shell in the 40mm family of ammunition. Its programmable fuze optimizes the effect against the target and increases the tactical flexibility of the 40mm weapon system by communication with the fire control system via a proximity fuze programmer. The 12.7mm machine gun (M2HB-QCB), of which the ship has two, is capable of firing explosive incendiary rounds and a full range of ball, tracer, armour-piercing and incendiary cartridges. In service around the world, it is the ideal weapon to mount on soft-skinned and armoured vehicles and also on patrol boats as the main or secondary armament.

SPECIFICATIONS

Type:	offshore patrol boat
Crew:	24
Displacement:	461 tonnes (454 tons)
Length:	54.5m (178.8ft)
Beam:	8m (26.24ft)
Draught:	2.5m (8.2ft)
Speed:	24.5 knots
Range:	6758km (4200 miles)
Missiles:	none
Guns:	1 x 40mm, 1 x 20mm, 2 x 12.7mm
Torpedoes:	none
Helicopters:	none
Aircraft:	none
Air Search Radar:	none
Surface Radar:	Decca 1226
Fire Control Radar:	none
Propulsion:	2 x diesels, 8000shp

SUFFREN

The Suffren class was the first French destroyer designed from the outset to carry surface-to-air missiles (SAMs). Three ships were planned with more to follow, but budget cutbacks reduced this number down to two. They have been significantly modernized over the years and will serve well into the twenty-first century. They are primarily used as air defence ships for French carriers. Anti-aircraft destroyers in general are intended to provide protection for maritime forces against missiles and air threats. They contribute to the surveillance and control of air space and have the task of intervening in any zone within a group, according to the needs of the mission. In air engagements, initial target detection is usually provided by the long-range air search radar. This is a three-dimensional, electronically stabilized, computer-controlled radar, which includes an automatic detection and tracking (ADT) capability. Target data is transferred, automatically or manually, to the computer of the Naval Tactical Data System (NTDS). NTDS forms the heart of the combat system, tying together the various subsystems, which collects and processes information from ship sensors, and from off-ship sensors, via radio digital data links. From NTDS, air targets are then sent to the weapons for actual engagement.

SPECIFICATIONS

Type:	destroyer
Crew:	355
Displacement:	7020 tonnes (6910 tons)
Length:	157.6m (517ft)
Beam:	15.5m (50.85ft)
Draught:	6.1m (20ft)
Speed:	34 knots
Range:	8206km (5100 miles)
Missiles:	Exocet, Masurca, Malafon
Guns:	2 x 100mm, 4–6 x 20mm
Torpedoes:	none
Helicopters:	none
Aircraft:	none
Air Search Radar:	DRBI 23
Surface Radar:	DRBV 1 SA
Fire Control Radar:	2 x DRBR 51, 1 x DRBC 33A
Propulsion:	steam turbines, 72,500shp

BRANDENBURG

The Type 123 Brandenburg class frigates were ordered in June 1989 to replace the Hamburg class. The frigates are primarily tasked with anti-submarine operations, but also contribute to anti-air defence, the tactical command of group forces and surface operations. The ships form part of the Wilhelmshaven-based 6th Frigate Squadron. They are armed with two twin launchers for Exocet surface-to-surface missiles (SSMs). A Lockheed Martin Mk 41 Mod 3 vertical launch system for 16 NATO Sea Sparrow medium-range SAMs is fitted. The ships also have two 21-cell launchers for the RAM (Rolling Airframe) short-range SAM. The RAM missile has infrared guidance and a range of 9.5km (15.28 miles). The ships' main gun is the Otobreda 76mm/62 Mk 75, which has a firing rate of 85 rounds per minute and a range of 16km (25.74 miles) anti-surface and 12km (19.3 miles) anti-air. There are also two 20mm Rh 202 guns. Two twin 324mm Mk 32 torpedo tubes are fitted for the AlliantTechsystems Mk 46 active/-passive anti-submarine torpedo. The electronic warfare suite includes the FL 1800 S-II electronic support measures and countermeasures system, developed by Daimler Chrysler Aerospace. In addition, two Otobreda SCLAR decoy dispensers are fitted for chaff and infrared flares.

SPECIFICATIONS

Type:	frigate
Crew:	218
Displacement:	4775 tonnes (4700 tons)
Length:	138.9m (455.7ft)
Beam:	16.7m (54.79ft)
Draught:	6.3m (20.66ft)
Speed:	29 knots
Range:	6436km (4000 miles)
Missiles:	Exocet SSM, Sea Sparrow, RAM
Guns:	1 x 76mm, 2 x 20mm
Torpedoes:	Mk 32, Mk 46
Helicopters:	2 x Sea Lynx
Aircraft:	none
Air Search Radar:	Signaal LW 08
Surface Radar:	Signaal SMART, 3D
Fire Control Radar:	2 x Signaal STIR 180 trackers
Propulsion:	2 x gas turbines, 51,680shp

BREMEN

The *Bremen* is designed primarily for anti-surface warfare missions with strong anti-air and anti-submarine warfare capability. Her combat system integrates target acquisition, navigation, communications, signal processing and weapon control functions, and the combat system's central computer calculates and evaluates the target data and then allocates data to the weapon systems. The system carries out multiple target search and track, target prioritization and automatic engagement of weapons. The ship's point defence system is based on the medium-range Sea Sparrow and the short-range RAM (Rolling Airframe) missile. Sea Sparrow is launched from two Mk 29 eight-cell launchers mounted side by side at the fore of the ship, above and behind the Otobreda gun. The RAM (RIM-116A) is installed at the aft of the ship above the helicopter deck. The ship is equipped with a 76mm Otobreda anti-air and anti-surface gun, and two 20mm guns are installed port and starboard. The ship accommodates two Sea Lynx helicopters, which are equipped with AQS-18D dipping sonar and two torpedoes: either Mk 46 or DM4 type. The hangar provides space, facilities and equipment for the maintenance of the two helicopters, and the flight deck is large enough for landing a Sea King-type helicopter.

SPECIFICATIONS

Type:	frigate
Crew:	200
Displacement:	3658 tonnes (3600 tons)
Length:	130m (426.5ft)
Beam:	14.5m (47.57ft)
Draught:	4.26m (13.97ft)
Speed:	30 knots
Range:	6436km (4000 miles)
Missiles:	Harpoon, Sea Sparrow, RAM
Guns:	1 x 76mm, 2 x 20mm
Torpedoes:	Mk 32, Mk 46
Helicopters:	2 x Sea Lynx
Aircraft:	none
Air Search Radar:	Signaal DA 08
Surface Radar:	Signaal DA 08
Fire Control Radar:	Signaal WM 25, Signaal STIR
Propulsion:	2 x gas turbines, 50,000shp

LUTJENS

The *Lutjens* is one of the two American Charles F. Adams class guided missile destroyers constructed in the late 1950s and early 1960s which Germany purchased. Despite periodic modernizations, the class was retired in the early 1990s from the US Navy. Modernization with the New Threat Upgrade (NTU) package was considered for these ships but was terminated since modernization would not have been cost effective given their limited service lives remaining. They have suffered severe machinery problems and are overdue for replacement. Notwithstanding mechanical problems, their armament is good. For example, they are equipped with Harpoon surface-to-surface missiles (SSMs). Their low-level, sea-skimming cruise trajectory, active radar guidance and warhead design assure high survivability and effectiveness. The Harpoon missile and its launch control equipment provide the capability to interdict ships at ranges well beyond those of other missiles. Once targeting information is obtained and sent to the Harpoon missile, it is fired. Once fired, the missile flies to the target location, turns on its seeker, locates the target and strikes it without further action from the firing platform. This allows the firing platform to engage other threats instead of concentrating on one at a time.

SPECIFICATIONS

Type:	destroyer
Crew:	340
Displacement:	4572 tonnes (4500 tons)
Length:	133.2m (437ft)
Beam:	14.3m (46.91ft)
Draught:	4.5m (14.76ft)
Speed:	32 knots
Range:	7200km (4500 miles)
Missiles:	Harpoon SSM, RAM
Guns:	2 x 20mm
Torpedoes:	Mk 46
Helicopters:	none
Aircraft:	none
Air Search Radar:	Lockheed SPS 40, Hughes SPS 52
Surface Radar:	Raytheon/Sylvania SPS 10
Fire Control Radar:	2 x Raytheon SPG 51, SPQ 9
Propulsion:	2 x steam turbines, 70,000shp

HMS CAMPBELTOWN

The *Campbeltown* is one of the four Type 22 Batch 3 frigates in Royal Navy service. The Type 22 is significantly larger than its more recent cousin, the Type 23. A major difference is that the Type 22 can accommodate two Lynx helicopters whereas the Type 23 can only accommodate one. However, in times of peace it is usual for only one Lynx to be deployed. Given that the Lynx is a powerful anti-submarine aircraft, this gives the Type 22 a more powerful offensive role. On the negative side, however, is the fact that in both the Batch 2 and Batch 3 frigates, the Sea Wolf SAM system is mounted in launchers which means that the ship does not have 360-degree defence from air attacks. The Type 22 frigate is now relatively old but it has served the Royal Navy well, and it is likely that when the next three Type 23 frigates enter service, the remaining three Batch 2 frigates will be paid off. What replaces the Batch 3 ships is yet to be decided, but that batch still has some operational life left. The Batch 3 remains a very versatile ship. Its offensive weaponry is a match to the Type 23, and its extra helicopter capability is an advantage. However, the Type 22 does not have the same stealth characteristics of the Type 23 and so it is likely that the Type 22 will be of more use in an escort role rather than as an independent command.

SPECIFICATIONS

Type:	frigate
Crew:	259
Displacement:	4978 tonnes (4900 tons)
Length:	148.1m (485.89ft)
Beam:	14.8m (48.55ft)
Draught:	6.4m (20.99ft)
Speed:	30 knots
Range:	7200km (4500 miles)
Missiles:	Harpoon SSM, Sea Wolf SAM
Guns:	1 x 4.5in, 1 x Goalkeeper, 2 x 30mm
Torpedoes:	Marconi Stingray
Helicopters:	2 x Sea Lynx or EH101 Merlin
Aircraft:	none
Air Search Radar:	Marconi Type 967/968
Surface Radar:	Marconi Type 967/968
Fire Control Radar:	2 x Marconi Type 911
Propulsion:	2 x gas turbines, 37,540shp

HMS CATTISTOCK

The Hunt class was envisaged as a sophisticated minesweeper designed principally to work in coastal and shallow water areas, such as the approaches to the River Clyde, where many of Britain's nuclear submarines are based. The ships were designed with noise-reduction features to avoid accidentally triggering mines. The Hunt class combines the roles of minehunter and minesweeper. There are three ways the ships can neutralize mines: towing wire sweeps in order to sever moored mines, so they rise to the surface and can then be destroyed by gunfire; using acoustic or influence sweeps to trigger mines on the seabed; and identifying mines on the sonar and then using clearance divers to place charges on them. The Hunt class also carries the Remote Control Mine Disposal System (RCMDS). This is a small, unmanned, remote-controlled submarine that can destroy underwater mines using explosives. Although the Hunt class ships have been criticized for being costly and for performing poorly in rough weather, when they entered service they were the most advanced mine warfare vessels on the naval scene and the world's largest Glass Reinforced Plastic (GRP) warships. These vessels have also served in the waters off Northern Ireland, where they have undertaken counter-terrorist missions close inshore.

SPECIFICATIONS

Type:	*mine countermeasures vessel*
Crew:	*45*
Displacement:	*737 tonnes (725 tons)*
Length:	*60m (196.85ft)*
Beam:	*9.85m (32.31ft)*
Draught:	*2.2m (7.21ft)*
Speed:	*14 knots*
Range:	*4800km (3000 miles)*
Missiles:	*none*
Guns:	*1 x 30mm, 2 x 7.62mm*
Torpedoes:	*none*
Helicopters:	*none*
Aircraft:	*none*
Air Search Radar:	*none*
Surface Radar:	*Type 1007 I band*
Fire Control Radar:	*none*
Propulsion:	*2 x diesels*

HMS CORNWALL

HMS *Cornwall* is the lead vessel of the new Eight Frigate Squadron consisting of her sister ships HMS *Cumberland*, *Campbeltown* and *Chatham*. She was launched on 14 October 1985 and commissioned in Falmouth, Cornwall, on 23 April 1988. She was the first Type 22 Batch 3 frigate, designed to detect and destroy submarines at long range using bow mounted "active" and towed-array "passive" sonars, plus torpedoes fired from triple tubes or dropped from the ship's helicopter. She is the first British ship to be fitted with Harpoon anti-ship missiles (ASMs), giving her an extensive anti-surface capability. The Sea Wolf and Goalkeeper anti-missile systems give the ship excellent self-defence against all airborne targets, including sea-skimming missiles. There are two main differences between the Batch 2 and Batch 3 Type 22s. The first is that the Batch 3 has a 114mm (4.5in) gun. This was a direct consequence of experience in the Falklands War, where it was discovered that the earlier Batch 1 Type 22s were at a disadvantage because of the absence of a gun. The 114mm gun is the same as that found on the Type 23, and is capable of firing at a rate of 25 rounds per minute up to a range of 22km (13.75 miles). The second significant difference is that the Batch 3 is equipped with Harpoon rather than Exocet.

SPECIFICATIONS

Type:	frigate
Crew:	259
Displacement:	4978 tonnes (4900 tons)
Length:	148.1m (485.89ft)
Beam:	14.8m (48.55ft)
Draught:	6.4m (20.99ft)
Speed:	30 knots
Range:	7200km (4500 miles)
Missiles:	Harpoon SSM, Sea Wolf SAM
Guns:	1 x 4.5in, 1 x Goalkeeper, 2 x 30mm
Torpedoes:	Marconi Stingray
Helicopters:	2 x Lynx or EH 101 Merlin
Aircraft:	none
Air Search Radar:	Marconi Type 967/968
Surface Radar:	Marconi Type 967/968
Fire Control Radar:	2 x Marconi Type 911
Propulsion:	2 x gas turbines, 37,540shp

GREAT BRITAIN **121**

HMS EDINBURGH

HMS *Edinburgh* is one of the youngest of the Type 42 destroyers, having been launched at Cammell Laird, Birkenhead, on 14 April 1983 and commissioned on 17 December 1985. She is a Batch 3 "stretched" Type 42, being some 16m (52.49ft) longer than her older sisters. *Edinburgh* has a crew of 301, including 26 officers. Her completion was delayed by several months to take in modifications as a result of lessons learned in the Falklands War. The Type 42 destroyer is an air defence platform, protecting herself and her group against attacks by enemy aircraft and missiles. She is equipped with Sea Dart surface-to-air missiles (SAMs), a 4.5in gun, Stingray torpedoes and the Vulcan Phalanx system as a last line of defence. Her Lynx helicopter carries Sea Skua anti-ship missiles. She underwent her most recent refit in the mid-1990s, at Rosyth, just across from Edinburgh, and her first deployment was to the Gulf on Armilla patrol. 1997 saw her mainly around the UK coast, with appearances at Staff College Sea Days, Joint Maritime Course, the Perisher (submarine commanders' qualifying course) and visits to Leith and Esbjerg in Denmark. Part of the following year was spent in the South Atlantic, patrolling the Falklands, making numerous visits to South American ports, and exercising with local navies.

SPECIFICATIONS

Type:	destroyer
Crew:	301
Displacement:	4750 tonnes (4675 tons)
Length:	141.4m (463.91ft)
Beam:	14.9m (48.88ft)
Draught:	5.8m (19.02ft)
Speed:	30 knots
Range:	6400km (4000 miles)
Missiles:	Sea Dart SAM
Guns:	1 x 4.5in, 2 x 20mm, 2 x Phalanx
Torpedoes:	Stingray
Helicopters:	1 x Lynx
Aircraft:	none
Air Search Radar:	Marconi/Signaal Type 1022
Surface Radar:	Type 996 or Type 992R
Fire Control Radar:	2 x Marconi Type 909
Propulsion:	2 x gas turbines, 54,400shp

HMS ILLUSTRIOUS

The Invincible class aircraft carrier HMS *Illustrious* was built at the Swan Hunter Shipbuilders yard in Wallsend and was commissioned in 1982. Her role is to provide a command headquarters for the task group and to support the operations of short take-off and vertical landing aircraft and helicopters. The ship accommodates over 1000 crew, including 350 aircrew with 80 officers. She also has capacity for an additional 500 Royal Marines. HMS *Invincible* and *Illustrious* each have three Thales Nederland Goalkeeper close-in weapons system (CIWS). Goalkeeper's Gatling 30mm gun provides a maximum firing rate of 4200 rounds per minute with a range of 1500m (4921ft). All three carriers are also equipped with two 20mm guns, which have a maximum range of 2km (1.25 miles) and a firing rate of 1000 rounds per minute. The Invincible class is fitted with the Racal Type 675(2) jamming system and either a UAA(2) (*Invincible*, *Ark Royal*) or UAT(8) (*Illustrious*) electronic support measures system also supplied by Racal. *Invincible* and *Ark Royal* are also to be fitted with the UAT ESM. The ship's decoy system is the Royal Navy's Outfit DLJ with Sea Gnat. There are eight 130mm six-barrel launchers produced by Hunting Engineering. Chemring and Pains Wessex produce the Sea Gnat chaff and infrared decoys.

SPECIFICATIONS

Type:	aircraft carrier
Crew:	740 plus 430 aircrew
Displacement:	20,930 tonnes (20,600 tons)
Length:	209.1m (686ft)
Beam:	36m (118.11ft)
Draught:	8m (26.24ft)
Speed:	28 knots
Range:	11,200km (7000 miles)
Missiles:	none
Guns:	3 x Goalkeeper, 2 x 20mm
Torpedoes:	none
Helicopters:	9 x Sea King, 3 x Sea King AEW
Aircraft:	9 x Sea Harrier
Air Search Radar:	Marconi/Signaal Type 1022
Surface Radar:	Marconi Type 992R (R 07)
Fire Control Radar:	2 x Marconi Type 909 or 909(1)
Propulsion:	4 x gas turbines, 94,000shp

HMS INVINCIBLE

The Invincible class aircraft carrier supports nine Harrier aircraft (both the Royal Air Force GR7 Harrier II and the Royal Navy F/A2 Sea Harrier), nine Sea King HAS 6 anti-submarine warfare (ASW) helicopters and three Sea King 2 airborne early warning (AEW) helicopters. Landing trials with the Merlin HM.1 helicopter have taken place on the *Ark Royal*, which will be the first carrier to deploy the Merlin. The helicopter's primary roles are anti-surface ship and submarine warfare, tracking and surveillance, amphibious operations, and search and rescue missions. It will operate from Type 22 and Type 23 class frigates, Invincible class aircraft carriers and various amphibious warfare ships and land bases. The first Merlin entered service with the Royal Navy in December 1998, at Royal Naval Air Station Culdrose, where the first squadron of Merlins was formed in October 2001. In September 2000, Merlin began operational trials on the Type 23 frigate HMS *Lancaster*. The cockpit is equipped with six high-definition colour displays plus an optional pilot's mission display. Dual flight controls are provided for the pilot and co-pilot on all versions of the EH101. The crew of the naval anti-submarine warfare version of the helicopter also includes an observer and an acoustic surveillance systems operator.

SPECIFICATIONS

Type:	aircraft carrier
Crew:	740 plus 430 aircrew
Displacement:	20,930 tonnes (20,600 tons)
Length:	209.1m (686ft)
Beam:	36m (118.11ft)
Draught:	8m (26.24ft)
Speed:	28 knots
Range:	11,200km (7000 miles)
Missiles:	none
Guns:	3 x Goalkeeper, 2 x 20mm
Torpedoes:	none
Helicopters:	9 x Sea King, 3 x Sea King AEW
Aircraft:	9 x Harrier
Air Search Radar:	Marconi/Signaal Type 1022
Surface Radar:	Marconi Type 992R (R 07)
Fire Control Radar:	2 x Marconi Type 909 or 909(1)
Propulsion:	4 x gas turbines, 94,000shp

HMS LIVERPOOL

This Sheffield class guided missile destroyer serves with the 3rd Destroyer Squadron. One of its weapons is the Vulcan Phalanx, which automatically engages functions usually performed by separate, independent systems such as search, detection, threat evaluation, acquisition, track, firing, target destruction, kill assessment and cease fire. Phalanx production started in 1978 with orders for 23 US Navy and 14 Foreign Military Sales (FMS) systems. Phalanx is a point-defence, total-weapon system consisting of two 20mm gun mounts that provide a terminal defence against incoming air targets. This close-in weapons system (CIWS) will automatically engage incoming anti-ship missiles and high-speed, low-level aircraft that have penetrated the ship's primary defence envelope. As a unitized system, CIWS automatically performs search, detecting, tracking, threat evaluation, firing, and kill assessments of targets while providing for manual override. Each gun mount houses a fire control assembly and a gun subsystem. The fire control assembly is composed of a search radar for surveillance and detection of hostile targets and a track radar for aiming the gun while tracking a target. The unique closed-loop fire control system gives CIWS the capability to correct its aim to hit fast-moving targets.

SPECIFICATIONS

Type:	destroyer
Crew:	253
Displacement:	4166 tonnes (4100 tons)
Length:	125m (410.1ft)
Beam:	14.3m (46.91ft)
Draught:	5.8m (19.02ft)
Speed:	29 knots
Range:	6400km (4000 miles)
Missiles:	Sea Dart SAM
Guns:	1 x 4.5in, 2–4 x 20mm, 2 x Phalanx
Torpedoes:	2 x triple torpedo tubes
Helicopters:	1 x Lynx
Aircraft:	none
Air Search Radar:	Marconi/Signaal Type 1022
Surface Radar:	Plessey Type 996
Fire Control Radar:	2 x Marconi Type 909
Propulsion:	2 x gas turbines, 54,000shp

HMS MANCHESTER

The Royal Navy has 11 Type 42 destroyers, also known as the Sheffield and Manchester classes, these being in three batches. HMS *Manchester*, and her sisters HMS *Gloucester*, *Edinburgh* and *York* were completed between 1983 and 1985. The main differences between these Batch 3 ships and the previous Batch 1 and 2 of the class are that the Batch 3 vessels are 16m (52.4ft) longer, slightly faster and have a slightly lower weapons fit, in addition to a slight increase in crew. Scheduled to be replaced by the long-awaited Type 45 destroyers, they will remain in service until at least 2007. The name ship of the class, HMS *Sheffield*, was lost in the Falklands War, as was her sister ship HMS *Coventry*. The original role for the Type 42 destroyer was the area air defence of the carrier/landing fleet, with a secondary anti-submarine role. This role was quite ably filled by the massive Type 82 HMS *Bristol*, but it was clear that this ship was too expensive to be procured in sufficient numbers and to be crewed, and for its size was under armed and had disadvantages, such as lack of a helicopter, a close-range armament system and anti-submarine torpedoes. The Type 42 is a much-reduced "economy" design on the smallest possible hull, with maximum automation for the smallest possible crew, carrying the heaviest possible armament.

SPECIFICATIONS

Type:	destroyer
Crew:	301
Displacement:	4750 tonnes (4675 tons)
Length:	141.4m (463.91ft)
Beam:	14.9m (48.88ft)
Draught:	5.8m (19.02ft)
Speed:	30 knots
Range:	6400km (4000 miles)
Missiles:	Sea Dart SAM
Guns:	1 x 4.5in, 2 x 20mm, 2 x Phalanx
Torpedoes:	Stingray
Helicopters:	1 x Sea Lynx
Aircraft:	none
Air Search Radar:	Marconi/Signaal Type 1022
Surface Radar:	Type 996 or Type 992R
Fire Control Radar:	2 x Marconi Type 909
Propulsion:	2 x gas turbines, 54,400shp

HMS MARLBOROUGH

The Type 23 Duke class was originally designed for anti-submarine warfare (ASW), but the addition of the vertical-launched Sea Wolf point missile defence system and the Harpoon SSM has expanded its role to include anti-surface warfare. Harpoon is undergoing an upgrade programme to improve its capabilities. Harpoon Block II will provide accurate long-range guidance for coastal, littoral and blue water ship targets by incorporating the global positioning system/inertial navigation system (GPS/INS) from the Joint Direct Attack Munitions (JDAM) programme. The existing 227kg (500lb) blast warhead will deliver lethal firepower against targets which include coastal anti-surface missile sites and ships in port. For the anti-ship mission, the GPS/INS provides improved missile guidance to the target area. The accurate navigation solution allows target ship discrimination from a nearby land mass using shoreline data provided by the launch platform. These Block II improvements will maintain Harpoon's high-hit probability while offering a 90 percent improvement in the separation distance between the hostile threat and local shorelines. Harpoon Block II will be capable of deployment from all platforms which currently have the Harpoon missile system by using existing command and launch equipment.

SPECIFICATIONS

Type:	frigate
Crew:	185
Displacement:	4267 tonnes (4200 tons)
Length:	133m (436.35ft)
Beam:	16.1m (52.82ft)
Draught:	5.5m (18.04ft)
Speed:	28 knots
Range:	12,480km (7800 miles)
Missiles:	Harpoon SSM, Sea Wolf SAM
Guns:	1 x 4.5in, 2 x 30mm
Torpedoes:	4 x 324mm tubes
Helicopters:	1 x Sea Lynx or EH101 Merlin
Aircraft:	none
Air Search Radar:	Plessey Type 996 (I), 3D
Surface Radar:	Plessey Type 996 (I), 3D
Fire Control Radar:	2 x Marconi Type 911
Propulsion:	4 x diesels, 52,300shp

HMS MONTROSE

Another of the Type 23 Duke class frigates, *Montrose* is armed with a Vickers 114mm (4.5in) gun that has a range of 22km (13.75 miles) against surface and 6km (3.75 miles) against airborne targets, and two Oerlikon 30mm guns with a range of 10km (6.25 miles) against surface and 3.5km (2.18 miles) against airborne targets. The ship has four 324mm torpedo tubes carrying Stingray torpedoes. Stingray has a depth of 750m (2461ft) and a range of 11km (6.87 miles). The sonar system, which is mounted at the front of the torpedo's guidance and control section, comprises a low-noise array assembly, transmitter unit, receiver and two digital signal processing units. The dual active/passive sonar operates with multiple, selectable transmit and receive beams. The ship's command and control system is a fully distributed Ada system based on technology using Intel processors, INMOS T800 transputers and a dual fibre-optic network. The satellite communications system is the SCOT 1D. Countermeasures systems include four Sea Gnat decoys and a Type 182 towed torpedo decoy. The Sea Gnats are mounted on 130mm six-barrel launchers. Thales Defence UAF-1 ESM is fitted to the first seven ships and Thales Defence's UAT(1) to the rest. The ships also boast the Thales Defence's Scorpion jammer.

SPECIFICATIONS

Type:	frigate
Crew:	185
Displacement:	4267 tonnes (4200 tons)
Length:	133m (436.35ft)
Beam:	16.1m (52.82ft)
Draught:	5.5m (18.04ft)
Speed:	28 knots
Range:	12,480km (7800 miles)
Missiles:	Harpoon SSM, Sea Wolf SAM
Guns:	1 x 4.5in, 2 x 30mm
Torpedoes:	4 x 324mm tubes
Helicopters:	1 x Lynx or EH 101 Merlin
Aircraft:	none
Air Search Radar:	Plessey Type 996 (I), 3D
Surface Radar:	Plessey Type 996 (I), 3D
Fire Control Radar:	2 x Marconi Type 911
Propulsion:	4 x diesels, 52,300shp

HMS NORFOLK

At present the Type 23 frigates have Thomson Marconi Sonar Type 2050 medium-range bow-mounted active/passive search and attack sonar and Thomson Type 2031Z very low-frequency passive search towed array sonar. However, the latter is to be replaced by the Type 2087 low-frequency active sonar (LFAS). This is a variable depth low-frequency transmitter and a passive, towed reception array. A contract for the Type 2087 development has been awarded to Thomson Marconi Sonar. The Type 2087, which is due to enter service in 2006, will have a greater range with bistatic and intercept capability. HMS *Norfolk* has a diesel-electric and gas (CODLAG) system, which consists of two Rolls-Royce Spey SM1A 34000hp gas turbines and two GEC 1.5MW 4400hp electric motors. There are also four GEC-Alsthom Paxman Valenta 12 RP2000CZ 7000hp auxiliary diesels. Using the diesel-electric motors, the economical speed is 15 knots and the range is 12,480km (7800 miles). The Type 23 carries the Lynx ASW helicopter, which is due to be replaced by the EH101 Merlin helicopter which entered service in December 1998. The Merlin helicopter provides search-and-attack capability for ASW and surface surveillance and over-the-horizon targeting for anti-surface warfare.

SPECIFICATIONS

Type:	frigate
Crew:	185
Displacement:	4267 tonnes (4200 tons)
Length:	133m (436.35ft)
Beam:	16.1m (52.82ft)
Draught:	5.5m (18.04ft)
Speed:	28 knots
Range:	12,480km (7800 miles)
Missiles:	Harpoon SSM, Sea Wolf SAM
Guns:	1 x 4.5in, 2 x 30mm
Torpedoes:	4 x 324mm tubes
Helicopters:	1 x Sea Lynx or EH101 Merlin
Aircraft:	none
Air Search Radar:	Plessey Type 996 (I), 3D
Surface Radar:	Plessey Type 996 (I), 3D
Fire Control Radar:	2 x Marconi Type 911
Propulsion:	4 x diesels, 52,300shp

HMS OCEAN

The primary role of HMS *Ocean* is to achieve the rapid landing of an assault force by helicopter and landing craft. The ship carries a crew of 255, an aircrew of 206 and 480 Royal Marines (an extra 320 Royal Marines can be accommodated in a short-term emergency). The ship has capacity for 40 vehicles but is not designed to land heavy tanks. There are four LCVP Mk 5 vehicle/personnel landing craft on davits, and the ship has full facilities for 12 EH101 Merlin and six Lynx helicopters, and landing and refuelling facilities for Chinook helicopters. Twenty Sea Harriers could be carried but not supported. The weapon systems include four Oerlikon twin 30mm guns together with three Phalanx Mk 15 close-in weapon systems (CIWS). HMS *Ocean* is equipped with eight Sea Gnat radar reflection/infrared emitting decoys. Sea Gnat was developed under a NATO collaborative project involving the USA, Germany, Norway, Denmark and Great Britain for protection against anti-ship guided missiles. The electronic support measures system is the Royal Navy's UAT, which is a radar warning receiver and electronic surveillance system that provides targeting data and identification of hostile radar threats. Also fitted is the Thales Type 675(2) shipborne jammer, which has a range of 500km (312 miles).

SPECIFICATIONS

Type:	amphibious helicopter carrier
Crew:	255 + 206 aircrew + 480 Marines
Displacement:	22,108 tonnes (21,760 tons)
Length:	203m (666ft)
Beam:	34m (111.54ft)
Draught:	6.6m (21.65ft)
Speed:	18 knots
Range:	12,800km (8000 miles)
Missiles:	none
Guns:	3 x Phalanx, 4 x twin 30mm
Torpedoes:	none
Helicopters:	12 x EH101 Merlin, 6 x Sea Lynx
Aircraft:	20 x Sea Harrier if no helicopters
Air Search Radar:	Siemens Plessey Type 996
Surface Radar:	2 x Kelvin Hughes Type 1007
Fire Control Radar:	ADAWS 2000 combat data system
Propulsion:	2 x diesels, 47,000shp

HMS SANDOWN

Sandown class minehunters are built almost entirely of non-magnetic materials and are designed to resist high shock levels. Their manoeuvrability is controlled, either manually or automatically, by using the Ship Position Control System (SPCS). The Sandown class is equipped with two underwater PAP 104 Mark 5 Remote Control Mine Disposal Vehicles. The vehicle is controlled via a 2000m (6561ft) fibre-optic cable. A lighting system, low-light level black and white camera and a colour camera are fitted. The vehicle is also fitted with a high-resolution sonar. The sensor data is transmitted back to the operations control centre on the ship. The main payload is a 100kg (220lb) mine-disposal charge which can be replaced by a manipulator. Wire cutters are used to release moored mines from the column of water above the sea bed. The mine disposal vehicles can be deployed to a depth of 300m (984ft). The Batch 2 Sandown class ships have a more powerful crane for deployment and recovery of the remotely controlled vehicles. The ship is equipped with two Barricade lightweight decoy launchers, which are capable of dispensing infrared decoys and chaff in confusion, distraction and centroid seduction modes of operation. The sonar system is the Type 2093 variable depth sonar which is deployed from a well in the hull.

SPECIFICATIONS

Type:	minehunter
Crew:	34
Displacement:	492 tonnes (484 tons)
Length:	52.5m (172.24ft)
Beam:	10.5m (34.44ft)
Draught:	2.3m (7.54ft)
Speed:	13 knots
Range:	4800km (3000 miles)
Missiles:	none
Guns:	1 x 30mm
Torpedoes:	none
Helicopters:	none
Aircraft:	none
Air Search Radar:	none
Surface Radar:	none
Fire Control Radar:	none
Propulsion:	2 x diesels

HMS SCOTT

The hull of the *Scott* survey ship encompasses three equipment/cargo holds, each separated by watertight bulkheads. Additionally, within the hull is a near complete double hull attained through the use of wing, deep and double-bottom tanks throughout her length. Above the main deck of the vessel is an approximately 26m (75ft) enclosed shelter deck in which is found space for a Work class remotely operated vehicle (ROV), specialized machinery and various control, testing, storage and workshop spaces. At the bow end of the main deck is an open working deck where cable machinery and trenching plows can be stored. Above the main deck is an upper platform deck where various plow equipment and control stations are located. There is also room for stowage, workshop and control containers, as well as several hydraulic equipment and cargo handling cranes. Under the main deck aft are the ship's engine spaces. The engine consists of a pair of main propulsion diesels coupled through reverse and reduction gears to propellers turning in Kort nozzles. Also located in the machinery spaces aft are the ship's service generators, switchboards, auxiliary machinery and the engine control room. For added thrust, the vessel is fitted with a combination of fully azimuthing and tunnel thrusters.

SPECIFICATIONS

Type:	ocean survey ship
Crew:	63
Displacement:	13,716 tonnes (13,500 tons)
Length:	131.5m (431.43ft)
Beam:	21.5m (70.53ft)
Draught:	4.57m (15ft)
Speed:	17.5 knots
Range:	unknown
Missiles:	none
Guns:	2 x .5in machine guns
Torpedoes:	none
Helicopters:	none
Aircraft:	none
Air Search Radar:	none
Surface Radar:	NR 58 DGPS & NR 230 DGPS
Fire Control Radar:	none
Propulsion:	2 x diesels

RFA SIR GALAHAD

The roles of the Sir Lancelot class landing ships are to transport troops, vehicles and equipment. To do this the vessels have doors and ramps at the stern and bow and internal ramps between decks, making them truly "drive through". They also have a crane forward of the superstructure for on/off loading equipment. They can carry aircraft on the helicopter pad behind the main superstructure and on the vehicle deck. They also have facilities to carry ammunition, repair vehicles and equipment. These vessels also have the capability of offloading directly onto a beach. To allow this they have a shallow draught (they are virtually flat bottomed) and two large anchors at the stern which they drop at sea before beaching and later use to pull themselves back out to sea. The ships are also capable of carrying mexifloats which can support military vehicles. All vessels have an aft flight deck capable of handling Sea King or Lynx helicopters, or in the case of *Sir Bedivere* a Merlin. Helicopters (including Chinooks) can also land on the vehicle deck, with the exception of *Sir Bedivere*, but the Chinook can only land when the ship is in port. The present vessel was built to replace the previous vessel of the same name lost in the Falklands conflict in 1982. The new Alternative Landing Ship Logistic (ALSL) vessels will replace *Sir Galahad*.

SPECIFICATIONS

Type:	large landing ship
Crew:	49 (plus up to 537 troops)
Displacement:	31,496 tonnes (31,000 tons)
Length:	140.47m (460.85ft)
Beam:	20m (65.61ft)
Draught:	4.57m (14.99ft)
Speed:	18 knots
Range:	9600km (6000 miles)
Missiles:	none
Guns:	2 x 20mm
Torpedoes:	none
Helicopters:	none
Aircraft:	none
Air Search Radar:	none
Surface Radar:	none
Fire Control Radar:	none
Propulsion:	2 x diesels, 47,000shp

HMS SOMERSET

The Type 23 Duke class frigates at present carry Lynx helicopters. However, in the near future they will be replaced by Merlin, which has the capacity to carry up to four homing torpedoes such as the Stingray or Mark 11 depth bombs. The anti-surface version is able to carry a range of air-to-surface missiles including sea-skimming anti-ship missiles. There are optional gun positions through removable windows, the starboard cargo door and the port crew door. In the anti-ship surveillance and tracking role, Merlin uses its tactical surveillance and over-the-horizon targeting radar to identify the positions of hostile ships and relay the data to the allied command ship. The aircraft has a state-of-the-art, integrated mission system, which processes data from an extensive array of on-board sensors, giving it an independent capability to search for, locate and attack submarine targets. It is this autonomous capability which makes Merlin unique among ASW helicopters. The aircraft and its mission system are managed by two computer systems, linked by dual data buses. The cockpit is designed for operation by a single pilot, with the autopilot allowing for hands-off flight for most of the mission. Merlin is normally flown by a crew of three: pilot, observer and aircrewman, who can all access the management computers.

SPECIFICATIONS

Type:	frigate
Crew:	185
Displacement:	4267 tonnes (4200 tons)
Length:	133m (436.35ft)
Beam:	16.1m (52.82ft)
Draught:	5.5m (18.04ft)
Speed:	28 knots
Range:	12,480km (7800 miles)
Missiles:	Harpoon SSM, Sea Wolf SAM
Guns:	1 x 4.5in, 2 x 30mm
Torpedoes:	4 x 324mm tubes
Helicopters:	1 x Lynx or EH101 Merlin
Aircraft:	none
Air Search Radar:	Plessey Type 996 (I), 3D
Surface Radar:	Plessey Type 996 (I), 3D
Fire Control Radar:	2 x Marconi Type 911
Propulsion:	4 x diesels, 52,300shp

HMS SOUTHAMPTON

This Sheffield class guided missile destroyer (Type 42 Batch 1/2) has as its main gun the Vickers 4.5in Mk 8 gun. Since the early 1990s, naval surface fire support capabilities have been limited to these small-calibre guns, which lack adequate range, accuracy and lethality. Targeting and fire control are still done manually, and the navy acknowledges that the communications links between fire support ships and their customers are inadequate. A growing threat from sea-skimming anti-ship missiles is forcing fire support ships to operate at ever-increasing ranges from shore, further limiting the utility of guns. Notwithstanding the claims of manufacturers, the accuracy of naval gunfire depends on the accuracy with which the position of the firing ship has been fixed. Navigational aids, prominent terrain features or radar beacons emplaced on the shore may be used to compensate for this limitation. Bad weather and poor visibility make it difficult to determine the position of the ship by visual means and reduce the observer's opportunities for locating targets and adjusting fire. Bad weather also might force the ship out to sea. If the ship is firing while under way, the line of fire in relation to the frontline may change. This could cancel the fire mission, because the long-range probable errors may endanger friendly forces.

SPECIFICATIONS

Type:	destroyer
Crew:	253
Displacement:	4166 tonnes (4100 tons)
Length:	125m (410.1ft)
Beam:	14.3m (46.91ft)
Draught:	5.8m (19.02ft)
Speed:	29 knots
Range:	6400km (4000 miles)
Missiles:	Sea Dart SAM
Guns:	1 x 4.5in, 2–4 x 20mm, 2 x Phalanx
Torpedoes:	2 x triple torpedo tubes
Helicopters:	1 x Lynx
Aircraft:	none
Air Search Radar:	Marconi/Signaal Type 1022
Surface Radar:	Plessey Type 996
Fire Control Radar:	2 x Marconi Type 909
Propulsion:	2 x gas turbines, 54,000shp

VIRAAT

The carrier *Viraat* was originally HMS *Hermes*. She was built by Vickers-Armstrong, Barrow-in-Furness, and laid down on 21 June 1944 as one of the Centaur class light fleet carriers. She was not launched until 16 February 1953, being laid up a further four years awaiting completion. Completed on 18 November 1959, in 1971 she was recommissioned as a commando carrier, and then in the late 1970s as an interim V/STOL carrier. After serving as the flagship of the Royal Navy's task force during the 1982 Falklands War, she was stricken on 1 July 1985. The *Hermes* was sold to India on 19 April 1986, and after a major refit at Devonport before transfer, and renamed INS *Viraat* (R22), she was commissioned into the Indian Navy on 12 May 1987. The current air group includes 12 or 18 Sea Harrier V/STOL fighters and seven or eight Sea King or Kamov "Hormone" anti-submarine warfare (ASW) helicopters. In emergencies, the *Viraat* can operate up to 30 Harriers. She is due for retirement by 2010 following an extensive modernization programme that began in 1999. India continues her attempts to build or acquire additional aircraft carriers, and there are ongoing discussions about the possible transfer of the Russian ship *Gorshkov*, probably refitted as a full-deck V/STOL carrier.

SPECIFICATIONS

Type:	aircraft carrier
Crew:	1550
Displacement:	29,159 tonnes (28,700 tons)
Length:	208.8m (685.03ft)
Beam:	27.4m (89.89ft)
Draught:	8.7m (28.54ft)
Speed:	28 knots
Range:	unknown
Missiles:	Sea Cat SAM
Guns:	2 x 40mm, 2 x dual 30mm
Torpedoes:	none
Helicopters:	7 x Sea King
Aircraft:	12 x Sea Harrier
Air Search Radar:	Marconi Type 996
Surface Radar:	Plessey Type 994
Fire Control Radar:	2 x Plessey Type 904
Propulsion:	steam turbines, 76,000shp

GIUSEPPE GARIBALDI

The flagship of the Italian Navy, the *Giuseppe Garibaldi* carrier can carry out anti-submarine warfare, command and control of naval and aero-naval forces, area surveillance, convoy escort, commando transportation and fleet logistic support. The ship can accommodate up to 18 helicopters, for example the Agusta Sikorsky SH-3D Sea King or the Agusta Bell AB212. Alternatively, the ship can accommodate 16 AV-8B Harrier II aircraft, or a mix of helicopters and Harriers. Her long-range surface-to-surface missile (SSM) system, the MBDA Otomat, is installed on the gun decks at the stern of the ship, two launchers on the port and two on the starboard side. The missile has active radar homing, is armed with a 210kg (462lb) warhead and has a range of 120km (75 miles). The MBDA Albatros surface-to-air missile (SAM) system provides short-range point defence. The Albatros eight-cell launchers are installed on the roof decks at the forward and stern end of the main island. The system uses the Aspide missile, which has a semi-active radar seeker and a range of 14km (8.75 miles). Forty-eight Aspide missiles are carried. Fire control for the Albatros is provided by three AESN NA 30 radar/electro-optical directors, which include infrared camera and laser rangefinder as well as the Alenia RTN 30X fire control radar.

SPECIFICATIONS

Type:	aircraft carrier
Crew:	550 plus 230 aircrew
Displacement:	13,584 tonnes (13,370 tons)
Length:	180m (590.55ft)
Beam:	33.4m (109.5ft)
Draught:	6.7m (21.98ft)
Speed:	30 knots
Range:	11,200km (7000 miles)
Missiles:	Otomat SSM, Aspide SAM
Guns:	6 x 40mm
Torpedoes:	A290 or Mk 46
Helicopters:	18 x SH-3D Sea King
Aircraft:	16 x AV-8B Harrier
Air Search Radar:	SPS 768 (RAN 3L)
Surface Radar:	SPS 774 (RAN IOS)
Fire Control Radar:	3 x AESN NA 30, 3 x NA 21
Propulsion:	4 x gas turbines, 81,000shp

LERICI

The *Lerici*'s command system integrates the tactical data system with the major platform and operational systems. It includes the following features: control and operation of the suite of sensors and weapons; control of the main and auxiliary propulsion systems; and integrated control of all internal and external communications, including the transfer of messages and mine countermeasures data from the tactical consoles. The Integrated Ship Communications System has been developed by CEA Technologies and allows the management of all mine countermeasures activities, including mission planning, minehunting, mine disposal and post-mission analysis. If manoeuvring during minehunting operations is being automatically controlled, the system controls the auxiliary propulsion system via the minehunter autopilot. This provides the ship with auto-track and auto-hover for effective mine countermeasures operations. This class of ship has proved popular with other nations. For example, in 1986 South Korea began construction of the Swallow/Chebi class minehunter, which was based on the Lerici class. The Swallow class minehunter had new types of sonar and mine countermeasure equipment that was expected to improve the navy's capability to locate and to eliminate minefields in international shipping lanes during wartime.

SPECIFICATIONS

Type:	*minehunter*
Crew:	*50*
Displacement:	*683 tonnes (672 tons)*
Length:	*52.5m (172.24ft)*
Beam:	*9.87m (32.38ft)*
Draught:	*2.95m (9.67ft)*
Speed:	*15 knots*
Range:	*4000km (2500 miles)*
Missiles:	*none*
Guns:	*1 x 20mm*
Torpedoes:	*none*
Helicopters:	*none*
Aircraft:	*none*
Air Search Radar:	*none*
Surface Radar:	*none*
Fire Control Radar:	*none*
Propulsion:	*1 x diesel*

LIBECCIO

This Maestrale class frigate is a multirole warship. Essentially the Maestrales are enlarged and much improved Lupo designs. The most notable difference is the larger hangar housing two helicopters and the more modern weaponry and sensors. Each ship in the class is equipped with infrared decoy flares as protection against hostile missiles. These comprise the AMBL-2A long-range decoy-launcher, capable of firing parachute-suspended sub-munitions. It complements the shorter-range Mk 2 AMBL-1B system, which is suitable for smaller warships. Decoy rounds are contained in 34-round Type C "suitcases", and a delayed-action mechanism ensures that the cloud is co-located with the chaff cloud. Each IR decoy burns for 30 seconds at an altitude of 15m (49.21ft). Laid down in August 1979, *Libeccio* was launched in September 1981 and commissioned in February 1983. The ship's main gun installed on the bow deck is the Otobreda 127mm gun which fires 32kg (70.4lb) rounds at a rate of 45 rounds per minute. The range of the gun is more than 15km (9.37 miles) against surface targets and 7km (4.37 miles) against airborne targets. Two Otobreda 40mm twin anti-aircraft guns fire 0.96kg (2.11lb) shells at a rate of 300 rounds per minute to a range of 4km (2.5 miles) for airborne targets and to 12km (7.5 miles) for surface targets.

SPECIFICATIONS

Type:	*frigate*
Crew:	*232*
Displacement:	*3251 tonnes (3200 tons)*
Length:	*122.7m (402.55ft)*
Beam:	*12.9m (42.32ft)*
Draught:	*4.2m (13.77ft)*
Speed:	*32 knots*
Range:	*9000km (6000 miles)*
Missiles:	*Otomat SSM, Aspide SAM*
Guns:	*1 x 127mm, 4 x 40mm, 2 x 20mm*
Torpedoes:	*Mk 46, Whitehead A184*
Helicopters:	*2 x AB 212 ASW*
Aircraft:	*none*
Air Search Radar:	*Selenia SPS 774 (RAN 10S)*
Surface Radar:	*SMA SPS 702*
Fire Control Radar:	*Selenia SPG 75, 2 x Selenia SPG 74*
Propulsion:	*2 x diesels, 50,000shp*

LUIGI DURAND DE LA PENNE

The MM *Luigi Durand De La Penne* (D560) and the second of this class of ship, MM *Francesco Mimbelli* (D561), were commissioned in 1993. The ships were constructed at Fincantieri's Riva Trigoso shipyard. They were both built to replace older vessels, but the construction of two additional units has been cancelled. They are multirole warships able to perform anti-air defence for protecting task forces and convoys, anti-submarine and anti-surface warfare operations, assistance during landing operations and coastal bombardment. The IPN 20 command and control system gathers information from the ship's sensors and communications and data networks in order to compile, maintain and display the tactical situation. The ship's surface-to-surface missile (SSM) system is the Otomat Mk 2. Four twin launchers are installed on the missile deck amidship between the two radar masts. The missile uses mid-course guidance and active radar homing to approach the target at high subsonic speed in sea-skimming mode. The ship is also equipped with the SM-2MR Tartar GMLS Mark 13 missile system for defence against medium-range airborne targets. The SM-2MR has semi-active radar guidance and a range of 70km (43.75 miles). The ships are to be fitted with the MBDA Milas anti-submarine missile system.

SPECIFICATIONS

Type:	destroyer
Crew:	400
Displacement:	5486 tonnes (5400 tons)
Length:	147.7m (484.58ft)
Beam:	16.1m (52.82ft)
Draught:	5m (16.4ft)
Speed:	31.5 knots
Range:	11,200km (7000 miles)
Missiles:	Otomat SSM, Aspide SAM
Guns:	1 x 127mm, 3 x 76mm
Torpedoes:	Whitehead A 290
Helicopters:	2 x AB 212 ASW
Aircraft:	none
Air Search Radar:	Selenia SPS 768 (RAN 3L)
Surface Radar:	SMA SPS 702
Fire Control Radar:	4 x Selenia SPG 775, 2 x SPG 51D
Propulsion:	2 x diesels, 55,000shp

MAESTRALE

The lead ship in the Maestrale class of frigates, she is armed with the Otomat 2 over-the-horizon ship-to-ship missile, which has a range of 100–180km (62–112 miles). Due to influences from the curvature of the earth, the detection systems on the attacking ship itself are limited to direct use within approximately 40km (25 miles). Therefore, to operate and effect the over-the-horizon ship-to-ship missile, either a surface ship or a shipborne helicopter must take the forward position to act as a midway station to transmit pertinent data, relating target information and flight information on the missile back to the attacking ship. The helicopter simultaneously transmits pertinent commands from the firing ship or the midway point to the missile in flight, allowing the missile to hit the target with precision. Surface-to-air missiles (SAMs) consist of the Aspide, basically a licensed version of the American Sparrow, which is employed as both an air-to-air and surface-to-air missile, and in the latter role it is launched from both ships and ground platforms. The AIM-7E Sparrow entered service in 1962 and was widely used as a standard for other variants such as the Sky Flash (United Kingdom) and Aspide. At present there are eight ships in this class, all being launched between 1982 and 1985.

SPECIFICATIONS

Type:	frigate
Crew:	232
Displacement:	3251 tonnes (3200 tons)
Length:	122.7m (402.55ft)
Beam:	12.9m (42.32ft)
Draught:	4.2m (13.77ft)
Speed:	32 knots
Range:	9600km (6000 miles)
Missiles:	Otomat SSM, Aspide SAM
Guns:	1 x 127mm, 4 x 40mm, 2 x 20mm
Torpedoes:	Mk 46, Whitehead A184
Helicopters:	2 x AB 212 ASW
Aircraft:	none
Air Search Radar:	Selenia SPS 774 (RAN 10S)
Surface Radar:	SMA SPS 702
Fire Control Radar:	Selenia SPG 75, 2 x SPG 74
Propulsion:	2 x diesels, 50,000shp

MINERVA

This class was built to replace the older corvettes of the DeCristofaro and Albatross classes. While designated frigates and performing many of their jobs, these ships are perhaps best described as large corvettes which undertake general patrol duties. Fitted for but not installed with Teseo SSMs, these ships are used mainly for patrol and escort duties. The main gun is the Otobreda 76mm model. The company's Super Rapid Gun Mount is a multirole weapon intended as an anti-missile, anti-aircraft (AA) weapon system with anti-ship and naval gunfire support (NGS) capability. The gun in the AA role can use special pre-fragmented ammunition. For surface roles, special types of armour-piercing ammunition are available. New technologies are incorporated in order to enhance system effectiveness, and the company has developed the 76/62 family of ammunition to increase the performance of the 76/62 guns in terms of blast, fragmentation and perforation effects against air and surface targets. The maximum firing range of this projectile is 20km (12.5 miles) (instead of 16km [10 miles] as for standard 76/62 ammunition), without loss in accuracy and payload. In addition, full compatibility with the 76/62 gun mount's ballistics and mechanical feeding system is ensured. *Minerva* was launched in 1987.

SPECIFICATIONS

Type:	corvette
Crew:	122
Displacement:	1305 tonnes (1285 tons)
Length:	86.6m (284.12ft)
Beam:	10.5m (34.44ft)
Draught:	3.2m (10.49ft)
Speed:	24 knots
Range:	5600km (3500 miles)
Missiles:	Otomat SSM, Aspide SAM
Guns:	1 x 76mm
Torpedoes:	Honeywell Mk 46
Helicopters:	none
Aircraft:	none
Air Search Radar:	Selenia SPS 774 (RAN 10S)
Surface Radar:	Selenia SPS 774 (RAN 10S)
Fire Control Radar:	Selenia SPG 76 (RTN 30X)
Propulsion:	2 x diesels

ORSA

The *Orsa* is a Lupo class frigate. There were originally four in the class: *Lupo*, *Sagittario*, *Perseo* and *Orsa*. However, they were joined by those of the the Lupo (Artigliere) class of light frigates, which were built by Fincantieri at its Ancona and Riva Trigoso shipyards. They were initially built for Iraq, but delivery was cancelled following the United Nations embargoes against that country in 1990. New weapons and communications systems were fitted to meet the requirements of the Italian Navy. MM *Artigliere* (F582), MM *Aviere* (F583) and MM *Granitiere* (F585), built at Ancona, were commissioned in 1994, 1995 and 1996 respectively. The third-of-class ship, MM *Bersagliere* (F584), commissioned in 1995, was built at the Riva Trigosa shipyard in Genoa. The ship's surface search radar is the Selenia SPS 774 and the navigation radar is the SMA SPN 703, both operating at I band. The Alenia RAN 10S air search radar operates at E and F bands and has a range of 150km (240 miles). The ship has four fire control radars. Two I and J band Alenia SPG 70 fire control radars are long-range systems for use with the NA 21 fire control system for the missiles and main gun. The Alenia SPG 74 short-range fire control radar is used for the Dardo and the 40mm guns. A DE 1160B active search-and-attack sonar is also fitted.

SPECIFICATIONS

Type:	frigate
Crew:	194
Displacement:	2565 tonnes (2525 tons)
Length:	113.2m (371.39ft)
Beam:	11.3m (37.07ft)
Draught:	3.7m (12.13ft)
Speed:	35 knots
Range:	6800km (4250 miles)
Missiles:	Otomat SSM, Sea Sparrow SAM
Guns:	1 x 127mm, 4 x 40mm
Torpedoes:	Whitehead A224
Helicopters:	1 x AB 212 ASW
Aircraft:	none
Air Search Radar:	Selenia SPS 774 (RAN 10S)
Surface Radar:	SMA SPS 702, SMA SPQ2
Fire Control Radar:	2 x Selenia SPG 70, 2 x SPG 74
Propulsion:	2 x diesels, 50,000shp

SAGITTARIO

Another Lupo class frigate, the *Sagittario* is fitted with an electronic warfare suite includes a radar warning system and jammer. The decoy system consists of two SCLAR 20 tube 105mm decoy launchers. The SCLAR launcher deploys chaff, infrared decoys or illuminating flares in confusion, distraction and seduction modes. The engines drive two shafts with controlled pitch propellers. The turbines provide a maximum speed of 35 knots and the diesel engines provide a speed of 20 knots. She carries the AB 212 helicopter made by Agusta of Italy. It is a licence-built version of the UH-1N Iroquois or Huey multirole helicopter. Its "Twin-Pac" turboshaft installation consists of two turbines driving a single output shaft, either being able to drive the rotor if the other fails. The turbines are mounted outboard of the main rotor mast with the nozzles carried side-by-side in a single housing. The fuselage is of all-metal, semi-monocoque construction with a short nose, extensive glazing (including "chin" windows for look-down visibility) around the cockpit, and doors on both sides. The cockpit is open to the main cabin, which has large, rearward-sliding doors on both sides. The main cabin can carry troops in jump seats or litters in fold-down racks. Landing gear is parallel skids on tubular struts with a tail skid at the end of the tail boom.

SPECIFICATIONS

Type:	frigate
Crew:	194
Displacement:	2565 tonnes (2525 tons)
Length:	113.2m (371.39ft)
Beam:	11.3m (37.07ft)
Draught:	3.7m (12.13ft)
Speed:	35 knots
Range:	6800km (4250 miles)
Missiles:	Otomat SSM, Sea Sparrow SAM
Guns:	1 x 127mm, 4 x 40mm
Torpedoes:	Whitehead A224
Helicopters:	1 x AB 212 ASW
Aircraft:	none
Air Search Radar:	Selenia SPS 774 (RAN 10S)
Surface Radar:	SMA SPS 702, SMA SPQ2
Fire Control Radar:	2 x Selenia SPG 70, 2 x SPG 74
Propulsion:	2 x diesels, 50,000shp

VITTORIO VENETO

The *Vittorio Veneto* is a purpose-built helicopter cruiser that followed from the design of the Andrea Doria class of the 1950s. The addition of a second deck gave her a much greater helicopter capacity. A large central lift is set immediately aft of the superstructure and two sets of fin stabilizers make her a very steady helicopter platform. She has gone through two major refits in 1981 and 1984 which have upgraded her weaponry and radars. She was also refitted in the late 1980s. Six 40mm guns in twin turrets were added, and the Terrier missiles were removed and replaced by 40 Standard SM1 and 20 ASROC (Anti-Submarine Rocket) missiles. Two SPG 55C standard fire control systems were added, as were four Otomat SSM launchers. Her previous role of Italian flagship was handed over to the carrier *Giuseppe Garibaldi* in 1995. As well as carrying AB 212 helicopters, she has the capacity to field Sea Kings. She is expected to be decommissioned in 2005, though her life may be extended pending developments regarding new Italian carriers that are currently in production, one being the so-called "new major vessel". Building work on the new vessel, which will be named the *Andrea Doria*, began at Fincantieri's shipyards in Riva Trigoso and Muggiano in July 2001 and the ship will be delivered in 2007.

SPECIFICATIONS

Type:	*helicopter cruiser*
Crew:	*557*
Displacement:	*9652 tonnes (9500 tons)*
Length:	*179.6m (589.23ft)*
Beam:	*19.4m (63.64ft)*
Draught:	*6m (19.68ft)*
Speed:	*32 knots*
Range:	*8000km (5000 miles)*
Missiles:	*Otomat SSM, SM-1ER SAM*
Guns:	*8 x 76mm, 6 x 40mm*
Torpedoes:	*ASROC, Mk 46*
Helicopters:	*6 x AB 212 ASW*
Aircraft:	*none*
Air Search Radar:	*Selenia SPS 768 (RAN 3L)*
Surface Radar:	*SMA SPS 702*
Fire Control Radar:	*4 x Selenia SPG 70, 2 x SPG 74*
Propulsion:	*2 x steam turbines, 73,000shp*

JACOB VAN HEEMSKERCK

The *Jacob Van Heemskerck* is the missile version of the Kortenaer class, giving up ASW helicopter facilities in favour of an area-defence missile. It is armed with the Sea Sparrow missile that provides the capability of destroying hostile aircraft, anti-ship missiles, and airborne and surface missile platforms with surface-to-air missiles (SAMs). The ship can also be used to detect missile launchings by a surface vessel utilizing the NSSMS surveillance radar capability. The NSSMS consists of a Guided Missile Fire Control System (GMFCS) Mk 91 and a Guided Missile Launching System (GMLS) Mk 29. The GMFCS is a computer-operated fire control system that provides automatic acquisition and tracking of a designated target, generates launcher and missile orders, and in the automatic mode initiates the firing command when the target becomes engageable. Although most of the NSSMS operations are carried out under automatic or semi-automatic conditions, the GMFCS permits operator intervention and override at any time. The GMLS is a rapid-reaction, lightweight launching system that provides on-mount stowage and launch capability of up to eight missiles. The GMLS responds to launcher position commands, missile orders, and control commands issued by the GMFCS. The NSSMS employs Sea Sparrow III missiles.

SPECIFICATIONS

Type:	air defence frigate
Crew:	197
Displacement:	3810 tonnes (3750 tons)
Length:	130.5m (428.14ft)
Beam:	14.6m (47.9ft)
Draught:	4.3m (14.1ft)
Speed:	30 knots
Range:	7520km (4700 miles)
Missiles:	Harpoon SSM, Sea Sparrow SAM
Guns:	1 x Goalkeeper, 2 x 20mm
Torpedoes:	Mk 46
Helicopters:	none
Aircraft:	none
Air Search Radar:	Signaal LW 08
Surface Radar:	Signaal ZW 06
Fire Control Radar:	2 x STIR 240, STIR 180
Propulsion:	2 x gas turbines, 51,600shp

ADMIRAL PANTELEYEV

Designed primarily as an anti-submarine warfare platform, with a long cruising range and underway replenishment capabilities, the Udaloy class, to which *Admiral Panteleyev* belongs, reflects design changes that addressed the shortcomings of the previous Krivak programme: the lack of helicopter facilities, limited sonar capabilities and light air defences. The Udaloy has two helicopter hangars with doors that serve as a ramp to the flight deck. The Udaloy's air defence system consists of eight Klinok launchers, plus AK-630 and AK-100 gun mounts. Following Udaloy's commissioning, designers began developing an upgrade package in 1982 to provide more balanced capabilities. The Project 1155.1 Fregat II Class Large ASW Ships (NATO Codename Udaloy II) is intended to be the Russian counterpart to the US Arleigh Burke class ships. The Udaloy II is modified by the replacement of the SS-N-14 by the SS-N-22, reflecting a change in emphasis from ASW to anti-shipping. Other changes include an improved self-defence capability with the addition of the gun and SAM systems. Powered by modern gas turbine engines, it is equipped with more capable sonars, an integrated air defence fire control system, and a number of digital electronic systems based on state-of-the-art circuitry.

SPECIFICATIONS

Type:	guided missile destroyer
Crew:	296
Displacement:	8839 tonnes (8700 tons)
Length:	163.5m (536.41ft)
Beam:	19.3m (63.32ft)
Draught:	7.5m (24.6ft)
Speed:	30 knots
Range:	6400km (4000 miles)
Missiles:	SA-N-9 SAM, SS-N-22, SA-N-11
Guns:	2 x 100mm, 4 x 30mm
Torpedoes:	Type 53
Helicopters:	2 x Ka-27 Helix A
Aircraft:	none
Air Search Radar:	Strut Pair, Top Plate, 3D
Surface Radar:	3 x Palm Frond
Fire Control Radar:	2 x Eye Bowl
Propulsion:	4 x gas turbines, 60,000shp

SOVREMENNY

The Sovremenny class is designed to engage hostile ships by means of missile attack, and to provide warships and transport ships with protection against ship and air attack. Intended primarily for anti-ship operations, it was designed to compliment anti-submarine warfare (ASW) Udaloy destroyers. The ships are similar in size to the US Navy's AEGIS-equipped missile cruisers, and are armed with an anti-submarine helicopter, 48 air defence missiles, eight anti-ship missiles, torpedoes, mines, long-range guns and a comprehensive electronic warfare system. The first Sovremenny class destroyer was commissioned in 1985. The ship is equipped with the Moskit anti-ship missile system with two quadruple launchers installed port and starboard of the forward island and set at an angle of about 15 degrees to the horizontal. The air defence system is the Shtil surface-to-air missile (SAM). Two Shtil systems are installed, each on the raised deck behind the two-barrelled 130mm guns fore and aft of the two islands. The ships are further equipped with 130mm guns and four six-barrel AK-630 artillery systems for defence. Anti-submarine defence is provided by two double 533mm torpedo tubes installed port and starboard, and two six-barrel anti-submarine rocket launchers, RBU-1000, with 48 rockets, and a Ka-27 or Ka-25 helicopter.

SPECIFICATIONS

Type:	guided missile destroyer
Crew:	344
Displacement:	8067 tonnes (7940 tons)
Length:	156m (511.81ft)
Beam:	17.3m (56.75ft)
Draught:	7.85m (25.75ft)
Speed:	32 knots
Range:	22,400km (14,000 miles)
Missiles:	SS-N-22 SSM, SA-N-7 SAM
Guns:	4 x 130mm, 6 x 30mm AA
Torpedoes:	Type 53
Helicopters:	1 x Ka-25B or 1 x Ka-27
Aircraft:	none
Air Search Radar:	Top Steer, Plate Steer, Top Plate
Surface Radar:	3 x Palm Frond
Fire Control Radar:	Front Dome, Kite Screech, Bass Tilt
Propulsion:	2 x steam turbines, 100,000shp

KUZNETSOV

This aircraft carrier supports strategic missile-carrying submarines, surface ships and maritime missile-carrying aircraft of the fleet. She is capable of engaging surface, subsurface and airborne targets. Superficially similar to American carriers, the design is essentially "defensive": supporting nuclear submarine bases. The lack of catapults precludes launching aircraft with heavy strike loads, and the air superiority orientation of the air wing is apparent. Originally designated Black-Com-2 class, then subsequently the Kremlin class, these ships were finally redesignated Kuznetsov class. Initially Western analysts anticipated that they would have a combined nuclear and steam (CONAS) propulsion plant similar to the *Kirov* battle cruiser. However, the class was in fact conventionally propelled with oil-fired boilers. The first unit was originally named *Tbilisi*, and subsequently renamed *Admiral Flota Svetskogo Soyuza Kuznetsov*. The *Admiral Kuznetsov* is now the only operational aircraft carrier in the Russian Navy. A variety of aircraft were tested on her. The first specially configured Su-25UT Frogfoot B, Su-27 Flanker and MiG-29 Fulcrum conventional jets landed on her deck in November 1989, aided by arresting gear. The Mig-29K completed test flights from the deck of the aircraft carrier, but was not selected for production.

SPECIFICATIONS

Type:	aircraft carrier
Crew:	1960 + 626 air wing + 40 flag
Displacement:	68,580 tonnes (67,500 tons)
Length:	280m (918.63ft)
Beam:	35.4m (116.14ft)
Draught:	10.5m (34.44ft)
Speed:	30 knots
Range:	unknown
Missiles:	SS-N-19 SSM, SA-N-9 SAM
Guns:	8 x CADS-N-1, 6 x 30mm
Torpedoes:	none
Helicopters:	15 x Ka-27, 3 x Ka-29 AEW
Aircraft:	20 x Su-27K, 4 x Su-25
Air Search Radar:	Sky Watch, 3D
Surface Radar:	Top Plate, 2 x Strut Pair
Fire Control Radar:	4 x Cross Sword
Propulsion:	2 x gas turbines, 200,000shp

PYTLIVYY

The Krivak I frigate was an entirely new design, initially believed in the West to be designed for offensive surface warfare. In reality, the class was intended primarily as a defensive ASW ship. The Krivak was designed as a less expensive and capable counterpart to the larger Kresta II and Kara classes, with which it originally shared the BPK designation. Sources all agree that the Krivak I series consisted of 19 units, with the first unit being completed in 1970, though there is a difference of views as to whether the final unit was completed in 1980 or 1982. And while there is agreement on the names of these units, there is further argument regarding construction chronology. For example, some sources suggest that the *Pytlivyy* was one of the earlier ships, completed in 1974, while others state that is was the last ship, completed in 1982. At least two, possibly three, Krivak I frigates were modernized between 1987 and 1994. Known in the West as MOD Krivak, different sources suggest that the Soviet designation was either Project No 1135.2 Mod or Project No 1135.6. This modification featured a new surface-to-surface missile (SSM) in place of the old ASW rocket launcher, along with improved electronics and sonars. It was planned for all Krivak-I units, but was scrapped on financial grounds.

SPECIFICATIONS

Type:	frigate
Crew:	200
Displacement:	3658 tonnes (3600 tons)
Length:	123.5m (405.18ft)
Beam:	14.3m (46.91ft)
Draught:	5m (16.4ft)
Speed:	32 knots
Range:	7360km (4600 miles)
Missiles:	SS-N-25, SA-N-4, SS-N-14
Guns:	2 x 100mm
Torpedoes:	Type 53
Helicopters:	1 x Ka-27 (on Krivak III)
Aircraft:	none
Air Search Radar:	Head Net, Top Plate, 3D
Surface Radar:	Don Kay, Palm Frond
Fire Control Radar:	2 x Eye Bowl, 2 x Owl Screech
Propulsion:	2 x gas turbines, 48,000shp

BALEARES

These ships are modified from the basic USS *Knox* class design with Spanish-built electronics and weapon systems. Where the aft helicopter facilities sit on US-built ships, this ship has an enhanced outfit of ASW weaponry. The guided missile frigates bring an anti-air warfare (AAW) capability to the frigate mission, but they have some limitations. Designed as cost-effective surface combatants, they lack the multi-mission capability necessary for modern surface combatants faced with multiple, high technology threats. The RUR-5 Anti-Submarine Rocket (ASROC) is a ballistic missile designed to deliver the Mk 46 Mod 5 torpedo to a water entry point. Navy surface ships employed the ASROC with two different payloads: either a nuclear depth charge or the Mk 44 or Mk 46 lightweight acoustic torpedo. The ASROC weapons were relatively small devices designed to fit inside the distinctive eight-cell box launcher found on almost all cruisers and destroyers. The torpedo is a very sophisticated weapon employing, for its time, state-of-the-art technology for the propulsion and guidance systems. The torpedo is about 2.43m (8ft) long, weighs about 273kg (600lb) and is also carried in tubes on escort ships. After water entry, the torpedo powers up and attacks a submarine using either passive or active sonar.

SPECIFICATIONS

Type:	frigate
Crew:	253
Displacement:	4244 tonnes (4177 tons)
Length:	133.6m (438.32ft)
Beam:	14.3m (46.91ft)
Draught:	4.7m (15.41ft)
Speed:	28 knots
Range:	7200km (4500 miles)
Missiles:	Harpoon, SM-1MR SAM
Guns:	1 x 127mm, 2 x 20mm
Torpedoes:	Mk 46 Mod 5
Helicopters:	none
Aircraft:	none
Air Search Radar:	Hughes SPS-52A, 3D
Surface Radar:	Raytheon SPS-10
Fire Control Radar:	SPG-53B, SPG-51 C, RAN 12 L
Propulsion:	4 x steam turbines, 35,000shp

PRINCIPE DE ASTURIAS

The *Principe de Asturias* aircraft carrier has been in service with the Spanish Navy since 1988. The layout of the ship was partly derived from the design of the US Navy's Sea Control Ship. It normally supports a maximum of 29 fixed- and rotary wing aircraft with up to 12 on deck and 13 aircraft in the hangar. In an emergency, though, a maximum of 17 aircraft can be stored in the hangar. The hangar deck itself is accessed by two flight deck lifts. The ship has four FABA Meroka Mod 2B close-in weapon systems (CIWS), with twelve-barrelled Oerlikon L120 20mm guns. The guns have a rate of fire of 1440 rounds per minute and a range of up to 2000m (6562ft). They are installed as follows: two on the stern deck and one each on the port and starboard side of the flight deck. The Meroka fire control system has tracking radar and an ENOSA thermal imager. A Mk 13 Mod 4 eight-cell launcher for the Harpoon anti-ship missile is also fitted. The ship's decoy system is the Super Rapid Bloom Offboard Countermeasures, Super RBOC. The six-barrel launchers can fire both chaff and flare cartridges up to a range of 4km (2.5 miles). The towed torpedo decoy is the SLQ-25 Nixie decoy. A hull noise and propeller suppression system reduces the ship's acoustic signature and increases the effectiveness of the acoustic countermeasures deployed.

SPECIFICATIONS

Type:	aircraft carrier
Crew:	600 plus 230 air wing
Displacement:	17,463 tonnes (17,188 tons)
Length:	195.9m (314.63ft)
Beam:	24.3m (79.72ft)
Draught:	9.4m (30.83ft)
Speed:	26 knots
Range:	10,400km (6500 miles)
Missiles:	Harpoon ASM
Guns:	4 x Mod 2B CIWS, 2 x 37mm
Torpedoes:	none
Helicopters:	10 x Sea King, 4 x AB 212, 2 x SH-60B
Aircraft:	6–12 x AV-8B
Air Search Radar:	Hughes SPS-52 C/D 3D
Surface Radar:	ISC Cardion SPS-55
Fire Control Radar:	1 x RAN 12L, 4 x VPS 2, 1 x RTN 1
Propulsion:	2 x gas turbines, 46,400shp

USS ANTIETAM

Modern US Navy guided missile cruisers perform primarily in a Battle Force role. Due to their extensive combat capability, these ships have been designated as Battle Force Capable (BFC) units. These multi-mission ships are built to be employed in support of Carrier Battle Groups, Amphibious Assault Groups, as well as interdiction and escort missions. The Ticonderoga class, using the highly successful Spruance hull, was initially designated as a guided missile destroyer (DDG), but was redesignated as a guided missile cruiser (CG) on 1 January 1980. They were the first surface combatant ships equipped with the AEGIS Weapons System, the most sophisticated air defence system in the world. The AEGIS combat system in Ticonderoga class cruisers, and the upgrading of older cruisers, have increased the anti-aircraft warfare (AAW) capability of surface combatants. The heart of the AEGIS system is the AN/SPY-1A radar, which automatically detects and tracks air contacts to over 320km (200 miles). The AEGIS Weapons System is designed to defeat attacking missiles and provide quick reaction, high firepower and jamming resistance to deal with the AAW threat expected to be faced by the Battle Group. *Antietam* was commissioned on 6 June 1987 and will be decommissioned in 2022.

SPECIFICATIONS

Type:	guided missile cruiser
Crew:	344
Displacement:	10,116 tonnes (9957 tons)
Length:	172.8m (566.92ft)
Beam:	16.8m (55.11ft)
Draught:	9.5m (31.16ft)
Speed:	30 knots
Range:	9600km (6000 miles)
Missiles:	SLCM, Harpoon ASM, SM-2 SAM
Guns:	2 x 127mm, 2 x Phalanx, 2 x 25mm
Torpedoes:	Mk 46, Mk 50
Helicopters:	2 x SH-60B
Aircraft:	none
Air Search Radar:	SPS-49
Surface Radar:	SPS-55
Fire Control Radar:	SPQ-9
Propulsion:	4 x gas turbines, 80,000shp

USS ARLEIGH BURKE

The *Arleigh Burke* was the first US Navy ship designed to incorporate shaping techniques to reduce radar cross-section to defeat enemy weapons and sensors. The ship is used in high-threat areas to conduct anti-air, anti-submarine, anti-surface and strike operations. All ships of this class have the AEGIS (the Greek word for shield) air defence system with the SPY-1D phased array radar. They are armed with a 90-cell Vertical Launching System capable of storing and rapidly firing a mix of Standard, Tomahawk and Vertically Launched ASROC (VLA) missiles. Other armament includes the Harpoon anti-ship missile, the 127mm (5in) gun with improvements that integrate it with the AEGIS Weapons System, and the Phalanx CIWS. AEGIS is designed to counter all current and projected missile threats. A conventional, mechanically rotating radar "sees" a target when the radar beam strikes that target once during each 360-degree rotation of the antenna. A separate tracking radar is then required to engage each target. In contrast, the computer-controlled AN/SPY-1D phased array radar of the AEGIS system brings all these functions together. The four fixed arrays send out beams of electromagnetic energy in all directions simultaneously, continuously providing a search and tracking capability for hundreds of targets at the same time.

SPECIFICATIONS

Type:	guided missile destroyer
Crew:	323
Displacement:	8433 tonnes (8300 tons)
Length:	153.8m (504.59ft)
Beam:	20.4m (66.92ft)
Draught:	9.44m (31ft)
Speed:	32 knots
Range:	7040km (4400 miles)
Missiles:	SLCM, Harpoon ASM, SM-2 SAM
Guns:	1 x 127mm, 2 x Phalanx
Torpedoes:	Mk 46, Mk 50
Helicopters:	none
Aircraft:	none
Air Search Radar:	SPY-1D 3D
Surface Radar:	SPS-67(V)3
Fire Control Radar:	3 x Raytheon/RCA SPG-62
Propulsion:	4 x gas turbines, 100,000shp

USS BARRY

The USS *Barry* is the second ship of the Arleigh Burke class. It is equipped with Tomahawk missiles, which are launched from her Mark 41 Vertical Launch System (VLS) – a multi-warfare missile launching system capable of firing a mix of missiles against airborne and surface threats. It is modular in design, with eight modules symmetrically grouped to form a launcher magazine. The modules contain all the necessary components for launching functions when interfaced with the AEGIS Weapons System. The 127mm gun, in conjunction with the Mark 34 Gun Weapon System, is an anti-ship weapon which can also be used for close-in air contacts or to support forces ashore with naval gunfire support. The AN/SQQ-89 integrated ASW suite is a very advanced anti-submarine warfare system. The AN/SQR-19 Tactical Towed Array Sonar (TACTAS) provides extremely long-range passive detection of enemy submarines, and the AN/SQS-53C hull-mounted sonar is used to actively and passively detect and locate submarine contacts. The ship also has the capability to land the SH-60B LAMPS Mark III helicopter, which can link to the ship for support in anti-submarine operations. These systems are supplemented by the SLQ-32V(2) electronic warfare suite, which includes passive detection systems and decoy countermeasures.

SPECIFICATIONS

Type:	guided missile destroyer
Crew:	323
Displacement:	8433 tonnes (8300 tons)
Length:	153.8m (504.59ft)
Beam:	20.4m (66.92ft)
Draught:	9.44m (31ft)
Speed:	32 knots
Range:	7040km (4400 miles)
Missiles:	SLCM, Harpoon ASM, SM-2 SAM
Guns:	1 x 127mm, 2 x Phalanx
Torpedoes:	Mk 46, Mk 50
Helicopters:	none
Aircraft:	none
Air Search Radar:	SPY-1D 3D
Surface Radar:	SPS-67 (V) 3
Fire Control Radar:	3 x Raytheon/RCA SPG-62
Propulsion:	4 x gas turbines, 100,000shp

USS BLUE RIDGE

USS *Blue Ridge* (LCC-19) was conceived in 1963 and was in the planning and design stage for four years. She was built by the Philadelphia Naval Shipyard in 1967 and commissioned on 14 November 1970. The *Blue Ridge* is the most capable command ship ever built, with an extremely sophisticated command and control system. The Joint Maritime Command Information System (JMCIS) consists of numerous powerful computers distributed throughout the ship, from which information and data from worldwide sources are entered into a central database. This single integrated database concentrates the available information into a complete tactical picture of air, surface and subsurface contacts, enabling the fleet commander to quickly assess and concentrate on any situation which might arise. This ability gives *Blue Ridge* a global command and control capability unparalleled in naval history. In addition to her sophisticated command and control system, an extremely refined communications system is also an integral part of the ship's design. Through an automated patch panel and computer-controlled switching matrix, any combination of communications equipment desired may be quickly connected. The "clean" topside area is intended to keep the ship's interference with her own communications system to a minimum.

SPECIFICATIONS

Type:	amphibious command ship
Crew:	842
Displacement:	18,944 tonnes (18,646 tons)
Length:	194m (636.48ft)
Beam:	32.9m (107.93ft)
Draught:	8.8m (28.87ft)
Speed:	23 knots
Range:	20,800km (13,000 miles)
Missiles:	GMLS Mk 25 Mod 1 SAM
Guns:	4 x 76mm, 2 x Phalanx
Torpedoes:	none
Helicopters:	room for 1 utility helicopter
Aircraft:	none
Air Search Radar:	SPS-48C, 3D, SPS-40C, Mk 23 TAS
Surface Radar:	SPS-65(V)1
Fire Control Radar:	2 x Mk 51
Propulsion:	1 x geared turbine

USS BOXER

Boxer is a Wasp class Landing Helicopter Dock (LHD), which are the largest amphibious ships in the world. Wasp class ships are the first to be specifically designed to accommodate the AV-8B Harrier jet and the LCAC hovercraft, along with the full range of US Navy and US Marine helicopters, conventional landing craft and amphibious assault vehicles to support a Marine Expeditionary Unit (MEU) of 2000 US Marines. They provide command and control and aircraft facilities for sea control missions, while operating with a Carrier Battle Group. They transport and land ashore not only troops, but also the tanks, trucks, jeeps, other vehicles, artillery, ammunition and various supplies necessary to support the amphibious assault mission. Monorail trains transport cargo and supplies from storage and staging areas throughout the ship to a well deck which opens to the sea through huge gates in the stern. There, the cargo, troops and vehicles are loaded aboard landing craft for transit to the beach. The well deck can be ballasted down for conventional craft to float out on their way to the assault area. Helicopter flights also transfer troops and equipment to the beach, while the ship's air traffic control capability simultaneously directs close air tactical support provided by aircraft and helicopter gunships.

SPECIFICATIONS

Type:	amphibious ship
Crew:	1108 plus 1894 US Marines
Displacement:	41,180 tonnes (40,532 tons)
Length:	257.3m (844.16ft)
Beam:	42.7m (140ft)
Draught:	8.1m (26.57ft)
Speed:	22 knots
Range:	15,200km (9500 miles)
Missiles:	Sea Sparrow SAM, RAM
Guns:	2 x Phalanx, 3 x 25mm
Torpedoes:	none
Helicopters:	42 x CH-46E Sea Knight
Aircraft:	6–8 x AV-8B Harrier
Air Search Radar:	SPS-48E, SPS-49 (V) 7
Surface Radar:	SPS-67
Fire Control Radar:	Mk 9
Propulsion:	2 x steam turbines, 70,000shp

UNITED STATES

USS JARRETT

This Oliver Hazard Perry class frigate was launched on 2 July 1983. The US Navy developed the FFG 7 class using the minimal manning concept, which means supervisors must accomplish all tasks with fewer people than larger ships. Below decks, two gas turbine engines provide power for propulsion that enables the ships to reach speeds in excess of 30 knots. These advanced propulsion units allow a ship to get underway quickly and rapidly change operating modes. The propulsion plant as well as the electrical power plant is computer controlled and monitored to ensure a smooth running and efficient system. The gas turbine engines can be started and be ready to come up to full power in five minutes. The Perry class was originally fitted with Raytheon's AN/SLQ-32(V)2, a self-defence electronic support measures (ESM) system offering limited frequency cover and questionable security. The SLQ-32 antennas in a Perry are carried at about 15.24m (50ft) above the waterline, providing an intercept range of only 37km (23 miles). Following the Iraqi Exocet attack on the USS *Stark* on 17 May 1987, it was decided to upgrade the (V)2 installation by adding a jammer codenamed "Sidekick". The new variant was later designated SLQ-32(V)5, and to date a number of (V)2s have been brought to the new standard.

SPECIFICATIONS

Type:	guided missile frigate
Crew:	216
Displacement:	4166 tonnes (4100 tons)
Length:	138.07m (453ft)
Beam:	14.32m (47ft)
Draught:	7.65m (22ft)
Speed:	30 knots
Range:	7200km (4500 miles)
Missiles:	Harpoon ASM, SM-1MR SAM
Guns:	1 x 76mm, 1 x Phalanx
Torpedoes:	Mk 46
Helicopters:	2 x SH-60B LAMPS
Aircraft:	none
Air Search Radar:	SPS-49
Surface Radar:	SPS-55
Fire Control Radar:	Mk 92
Propulsion:	2 x gas turbines, 40,000shp

USS JOHN C. STENNIS

The mission of USS *John C. Stennis* and her embarked air wing is to conduct sustained combat air operations while forward deployed in the global arena. Her two nuclear reactors give her virtually unlimited range and endurance and a top speed in excess of 30 knots. The ship's four catapults and four arresting gear engines enable her to launch and recover aircraft rapidly and simultaneously. The ship carries approximately three million gallons of fuel for her aircraft and escorts, and enough weapons and stores for extended operations without replenishment. USS *John C. Stennis* also has extensive repair capabilities, including a fully equipped aircraft intermediate maintenance department, a micro-miniature electronics repair shop, and numerous ship repair shops. The air wing can destroy enemy aircraft, ships, submarines and land targets, or lay mines hundreds of miles from the ship. Carriers usually operate in the centre of the battle group, with the surrounding ships providing outer defence. However, should aircraft or missiles penetrate this outer ring, USS *John C. Stennis* has Sea Sparrow short-range, surface-to-air missile (SAM) systems, the Phalanx close-in weapon system (CIWS) for cruise missile defence, and the SLQ-32 Electronic Warfare System. The ship cost $3.5 billion US, and her projected service life is 50 years.

SPECIFICATIONS

Type:	aircraft carrier
Crew:	3200 plus 2480 air wing
Displacement:	98,552 tonnes (97,000 tons)
Length:	332.84m (1092ft)
Beam:	78.33m (257ft)
Draught:	11.3m (37.07ft)
Speed:	30 knots
Range:	unlimited
Missiles:	Sea Sparrow SAM
Guns:	4 x Phalanx
Torpedoes:	none
Helicopters:	8 x SH-3 G/H or SH-60F
Aircraft:	20 x F-14, 20 x F/A-18, 4 x EA-6B, 16 x A-6E, 4 x E-2C, 6 x S-3A/B
Air Search Radar:	SPS-48E, SPS-49(V)5, Mk 23 TAS
Surface Radar:	SPS-67V
Fire Control Radar:	3 x Mk 91

USS JOHN F. KENNEDY

USS *John F. Kennedy* (CV 67) was named after the 35th President of the United States. The ship's keel was laid on 22 October 1964 at the Newport News Shipbuilding and Drydock Company in Virginia. President Kennedy's nine-year-old daughter Caroline christened the ship in May 1967 in a ceremony held at Newport News, and the ship subsequently entered naval service on 7 September 1968. *Kennedy* was originally designed as a CVA 67 attack aircraft carrier. In the early 1970s, however, the classification was changed to CV 67, indicating the ship was capable of supporting anti-submarine warfare aircraft, making it an all-purpose, multi-mission aircraft carrier. In September 1995, the USS *John F. Kennedy* became the Naval Reserve's first aircraft carrier. Homeported at Mayport, Florida, her primary function during contingency operations is to provide a surge capability, and in peacetime to support US Navy force training requirements. As with all other reserve ships, she remains fully mission ready. She completed a two-year comprehensive overhaul in the Philadelphia Naval Shipyard on 13 September 1995. Following the overhaul, she moved to her new homeport at the Mayport Naval Station in Mayport, Florida. She has also served as a test bed for the Cooperative Engagement Capability (CEC) programme.

SPECIFICATIONS

Type:	aircraft carrier
Crew:	3117 plus 2480 air wing
Displacement:	83,312 tonnes (82,000 tons)
Length:	320.6m (1051.83ft)
Beam:	76.8m (252ft)
Draught:	11.4m (37.4ft)
Speed:	32 knots
Range:	19,200km (12,000 miles)
Missiles:	Sea Sparrow SAM
Guns:	3 x Phalanx
Torpedoes:	none
Helicopters:	8 x SH-3 G/H or SH-60F
Aircraft:	20 x F-14, 20 x F/A-18, 4 x EA-6B, 16 x A-6E, 4 x E-2C
Air Search Radar:	SPS-48 C/E, SPS-49 (V) 5, Mk 23 TAS
Surface Radar:	SPS-10F
Fire Control Radar:	3 x Mk 91

USS KAUFFMAN

The *Kauffman* (FFG 59) is one of the so-called "long hull" Oliver Hazard Perry class frigates. Perry class ships were produced in two variants, known as "short hull" and "long hull", with the later variant being 2.43m (8ft) longer than the short hull version. The long-hull ships (FFG 8, 28, 29, 32, 33, 36–61) carry the SH-60B LAMPS III helicopters, while the short-hull units carry the less-capable SH-2G. A total of 55 FFG 7 Oliver Hazard Perry class ships were built, including 51 for the US Navy and four for the Royal Australian Navy. Of these, 33 are in active commissioned service and 10 are in the Naval Reserve Force (NRF). The "short hull" Perry class frigates are being retired at an advanced rate, even though they have 20 years left on their service life. The US Navy had hoped to phase out construction of this class with the 1983 ships, FFG 59 and 60, but the US Congress authorized (but did not fully fund) FFG 61 in 1984. The Naval Reserve currently operates 10 Oliver Hazard Perry class frigates. These ships maintain full readiness status and deploy with their Active Component counterparts when needed. One of their primary missions, which they fulfil simply by being available, is to make it possible for the Active Component to maintain its operating tempo at acceptable levels.

SPECIFICATIONS

Type:	*guided missile frigate*
Crew:	*216*
Displacement:	*4166 tonnes (4100 tons)*
Length:	*138.07m (453ft)*
Beam:	*13.71m (45ft)*
Draught:	*4.5m (14.76ft)*
Speed:	*30 knots*
Range:	*7200km (4500 miles)*
Missiles:	*Harpoon ASM, SM-1MR SAM*
Guns:	*1 x 76mm, 1 x Phalanx*
Torpedoes:	*Mk 46*
Helicopters:	*2 x SH-60B LAMPS*
Aircraft:	*none*
Air Search Radar:	*SPS-49*
Surface Radar:	*SPS-55*
Fire Control Radar:	*Mk 92*
Propulsion:	*2 x gas turbines, 41,000shp*

USNS KILAUEA

The ammunition ship's mission is the delivery of bombs, bullets, missiles, mines, projectiles, powder, torpedoes and various other explosive devices and incendiaries to the various ships in the fleet at sea. This type of support is necessary in order to achieve and maintain the US Navy's requirement for a high degree of logistical independence. The ships have four cargo holds, which break down into 14 magazines. A magazine is the level within the cargo hold, and is defined as a magazine due to the stowage of ammunition and the requisite fire-detecting and fire-fighting items found on each level. As well as their delivery of ordnance and other goods, these ships carry out a replenishment-at-sea capability for limited quantities of fuel, water and combat stores. The ships also have facilities for limited ship repair and maintenance services, as well as special project services. The four cargo holds are serviced by six high-speed cargo weapons elevators. The ships have a certified helicopter flight deck and can handle any US military helicopter as well as most commercial and allied helicopters. There are seven cargo transfer stations and one fuel delivery station. The ships can also receive fuel at sea from any of four stations, and there are four cargo booms which allow for shore or barge transfer.

SPECIFICATIONS

Type:	ammunition ship
Crew:	180
Displacement:	20,492 tonnes (20,169 tons)
Length:	171.9m (564ft)
Beam:	24.68m (81ft)
Draught:	9.44m (31ft)
Speed:	20 knots
Range:	unknown
Missiles:	none
Guns:	none
Torpedoes:	none
Helicopters:	2 x UH-46E
Aircraft:	none
Air Search Radar:	none
Surface Radar:	unknown
Fire Control Radar:	none
Propulsion:	1 x turbine, 22,000shp

USS KITTY HAWK

The USS *Kitty Hawk* is a conventionally powered aircraft carrier. Combined with the aircraft of Carrier Air Wing Five, it carries F-14, F/A-18, EA-6B, S-3 A/B, E-2CA aircraft and SH-3 or SH-60 helicopters, which gives it a multi-dimensional response to air, surface and subsurface threats. *Kitty Hawk* underwent two overhauls in the Bremerton, Wash., Naval Shipyard in 1977 and 1982. The ship's most significant maintenance period, however, was a Service Life Extension Program (SLEP) at the Philadelphia Naval Shipyard beginning from 1987 through 1991. That rigorous four-year overhaul added an estimated 20 years to the planned 30-year life of the ship. Over a three-month period in early 1998, nearly 4000 shipyard workers, sailors and contractors completed $65 million US in repairs (involving over 500 major jobs) in the Complex Overhaul of the dry-docked *Kitty Hawk*. All four of the *Hawk*'s screws were repaired (number three was replaced), and all the line shaft bearings were replaced. Containments were built around the shafts to maintain temperature and humidity levels while complex fibreglass work was completed. For the rudders, large holes were cut through the decks, and the rudders and all associated systems were removed. Refurbished rudders were then re-machined and installed.

SPECIFICATIONS

Type:	aircraft carrier
Crew:	3117 plus 2480 air wing
Displacement:	87,376 tonnes (86,000 tons)
Length:	325.83m (1069ft)
Beam:	76.8m (252ft)
Draught:	11.4m (37.4ft)
Speed:	32 knots
Range:	19,200km (12,000 miles)
Missiles:	Sea Sparrow SAM
Guns:	3 x Phalanx
Torpedoes:	none
Helicopters:	8 x SH-3 G/H or SH-60F
Aircraft:	20 x F-l4, 20 x F/A-18, 6 x S-3A/B 4 x EA-6B, 16 x A-6E, 4 x E-2C
Air Search Radar:	SPS-48 C/E, SPS-49 (V) 5, Mk 23 TAS
Surface Radar:	SPS-10F
Fire Control Radar:	3 x Mk 91

USS LAKE ERIE

At the heart of this Ticonderoga class guided missile cruiser is the AEGIS combat system, which integrates electronic detection, command and decision programmes, and engagement systems. *Lake Erie*'s computer-controlled phased array radar eliminates the need for separate search and track radars by simultaneously performing both functions. The four fixed SPY radar arrays form small steerable beams of electromagnetic energy that provide almost instantaneous full radar coverage, capable of tracking hundreds of contacts at the same time. The engineering system aboard *Lake Erie* represents the latest technology in warship construction. Four LM-2500 gas turbine engines supply the ship with power, and are very similar to the engines that power the larger airliners used throughout the world. With all four engines on-line, 80,000 shaft horsepower is available to propel the ship. The Controlled Reversible Pitch Propellers optimize the ship's speed and manoeuvrability through the water. By varying the pitch and the revolutions per minute of the screws the ship can go from full ahead to a complete stop in two ship lengths. Electricity is supplied by three gas turbine generators, each one providing 2500kW of power to sustain the ship's functions. During operations, two generators are normally on-line.

SPECIFICATIONS

Type:	*guided missile cruiser*
Crew:	*344*
Displacement:	*10,116 tonnes (9957 tons)*
Length:	*172.8m (566.92ft)*
Beam:	*16.8m (55.11ft)*
Draught:	*9.5m (31.16ft)*
Speed:	*30 knots*
Range:	*9600km (6000 miles)*
Missiles:	*SLCM, Harpoon SSM, SM-2 SAM*
Guns:	*2 x 127mm, 2 x Phalanx, 2 x 25mm*
Torpedoes:	*Mk 46, Mk 50*
Helicopters:	*2 x SH-60B*
Aircraft:	*none*
Air Search Radar:	*SPS-49(V)8*
Surface Radar:	*SPS-55*
Fire Control Radar:	*SPQ-9*
Propulsion:	*4 x gas turbines, 80,000shp*

USS NASSAU

The primary war-fighting mission of the LHA 1 (Landing Helicopter Assault 1) Tarawa class is to land and sustain US Marines on any shore during hostilities. The ships serve as the centrepiece of a multi-ship Amphibious Readiness Group (ARG). Some 3000 sailors and US Marines contribute to a forward-deployed ARG composed of approximately 5000 personnel. The ships maintain what the US Marine Corps calls "tactical integrity": getting a balanced force to the same place at the same time. One LHA can carry a complete US Marine battalion, along with the supplies and equipment needed in an assault, and land them ashore by either helicopter or amphibious craft. This two-pronged capability, with the emphasis on airborne landing of troops and equipment, enables the US Navy and Marine Corps to fulfil their mission. Whether the landing force is involved in an armed conflict, acting as a deterrent force in an unfavourable political situation or serving in a humanitarian mission, the class offers tactical versatility. The Tarawa class is designed to operate independently or as a unit of a force, as a flagship, or individual ship unit in both air and/or surface assaults. To allow them to do this the vessels incorporate the best design features and capabilities of several amphibious assault ships currently in service.

SPECIFICATIONS

Type:	amphibious assault ship
Crew:	964 plus 1900 US Marines
Displacement:	40,564 tonnes (39,925 tons)
Length:	254.2m (833ft)
Beam:	40.2m (131ft)
Draught:	7.9m (25.91ft)
Speed:	24 knots
Range:	16,000km (10,000 miles)
Missiles:	RAM
Guns:	2 x Phalanx
Torpedoes:	none
Helicopters:	9 x CH-53D, 12 x CH-46D/E
Aircraft:	6 x AV-8B
Air Search Radar:	SPS-40, SPS-48E
Surface Radar:	SPS-67
Fire Control Radar:	Mk 23 TAS
Propulsion:	2 x steam turbines, 70,000shp

USNS NIAGRA FALLS

The mission of the Mars class combat stores ship is to conduct underway replenishment in support of US Navy operating forces by providing refrigerated stores, dry provisions, technical spares, general stores, fleet freight, mail and personnel by alongside or vertical replenishment means. To do this they are fitted with state-of-the-art replenishment-at-sea equipment. The Mars class combat stores ships are augmented by three T-AFS 8 Sirius class stores ships purchased from Great Britain (built in England in 1965 and 1966, they were extensively modernized with improved communications and underway replacement facilities). The ships of Military Sealift Command's (MSC's) Naval Fleet Auxiliary Force (NFAF) are the lifeline to US Navy ships at sea. *Niagra Falls* is part of the MSC Far East in the Western Pacific and Indian Oceans; to provide a combat-ready logistics force, sustained sealift, and special mission ships as required in support of unified and fleet commanders. *Niagra Falls* was commissioned on 29 April 1967. As well as this ship there are two others of the class currently in service: the *Concord* was commissioned on 27 November 1968, and the *San Jose* was commissioned on 23 October 1970. All ships were built by the National Steel and Shipbuilding Company.

SPECIFICATIONS

Type:	combat stores ship
Crew:	176
Displacement:	17,659 tonnes (17,381 tons)
Length:	177m (581ft)
Beam:	24m (79ft)
Draught:	8.53m (28ft)
Speed:	20 knots
Range:	unknown
Missiles:	none
Guns:	2 x Phalanx
Torpedoes:	none
Helicopters:	2 x UH-46
Aircraft:	none
Air Search Radar:	unknown
Surface Radar:	SPS-67
Fire Control Radar:	Mk 23 TAS
Propulsion:	3 x steam turbines, 22,000shp

USS NIMITZ

The lead ship of her class, the *Nimitz* has four aircraft elevators that bring aircraft to the flight deck from the hangars below. Small tractors spot the planes on the flight deck. Aviation fuel is pumped up from tanks below, and bombs and rockets are brought up from the magazines. Powerful steam catapults, nicknamed "Fat Cats", can accelerate 37-ton jets from zero to a safe flight speed of up to 288km/h (180mph) in about 91.5m (300ft) and in less than three seconds. The weight of each aircraft determines the amount of thrust provided by the catapult. When landing, pilots use a system of lenses to guide the aircraft "down the slope": the correct glide path for landing. The four arresting wires, each consisting of 50.8mm (2in) thick wire cables connected to hydraulic rams below decks, drag landing aircraft going as fast as 240km/h (150mph) to a stop in less than 122m (400ft). The "Air Boss" and his staff coordinate the entire operation, which is carefully monitored from the flight deck level as well as by the captain on the bridge. The functions of the flight deck crew are identified by the colours they wear: yellow for officers and aircraft directors; purple for fuel handlers; green for catapult and arresting gear crews; blue for tractor drivers; brown for chock and chain runners; and red for crash and salvage teams and the ordnance handlers.

SPECIFICATIONS

Type:	aircraft carrier
Crew:	3200 plus 2480 air wing
Displacement:	96,520 tonnes (95,000 tons)
Length:	332.84m (1092ft)
Beam:	76.8m (252ft)
Draught:	11.3m (37ft.07)
Speed:	30 knots
Range:	unlimited
Missiles:	Sea Sparrow SAM
Guns:	3 x Phalanx
Torpedoes:	none
Helicopters:	8 x SH-3G/H or SH-60F
Aircraft:	20 x F-14, 20 x F/A-18, 4 x EA-6B, 16 x A-6E, 4 x E-2C, 6 x S-3 A/B
Air Search Radar:	SPS-48E, SPS-49(V)5
Surface Radar:	SPS-67V
Fire Control Radar:	3 x Mk 91

USS PRINCETON

USS *Princeton* (CG 59) is the US Navy's first cruiser equipped with the AN/SPY-1B radar system, which will provide a significant improvement in the detection capabilities of the AEGIS Weapons System. This radar system incorporates significant advances over earlier radars, particularly in its resistance to enemy electronic countermeasures (ECM). With the SPY-1B radar and the ship's Mk 99 fire control system, the ship can guide its Standard Missile to intercept hostile aircraft and missiles at extended ranges. Anti-ship cruise missile capability is provided by Harpoon missiles, capable of striking surface targets at ranges beyond 104km (65 miles). The AN/SQR-19 tactical towed array system provides long-range passive detection of enemy submarines, while the hull-mounted AN/SQS-53B sonar can be used to detect and localize submarine contacts. Two LAMPS Mk III multi-purpose helicopters function as extensions of the ship to assist in both submarine prosecution and surface surveillance and targeting. In addition, the AEGIS system will be capable of providing a threat-wide defence against tactical ballistic missiles. In addition to fulfilling its traditional missions, *Princeton* is equipped for strike warfare using the vertically-launched Tomahawk land-attack cruise missile.

SPECIFICATIONS

Type:	*guided missile cruiser*
Crew:	*344*
Displacement:	*10,116 tonnes (9957 tons)*
Length:	*172.8m (566.92ft)*
Beam:	*16.8m (55.11ft)*
Draught:	*9.5m (31.16ft)*
Speed:	*30 knots*
Range:	*9600km (6000 miles)*
Missiles:	*SLCM, Harpoon, SM-2MR M*
Guns:	*2 x 127mm, 2 x Phalanx, 2 x 25mm*
Torpedoes:	*Mk 46, Mk 50*
Helicopters:	*2 x SH-60B*
Aircraft:	*none*
Air Search Radar:	*SPS-49(V)8*
Surface Radar:	*SPS-55*
Fire Control Radar:	*SPQ-9*
Propulsion:	*4 x gas turbines, 80,000shp*

USS SENTRY

Avenger class ships are designed as minehunter-killers capable of finding, classifying and destroying moored and bottom mines. The mine countermeasures (MCM) ship performs precise navigation and clears minefields by sweeping moored, magnetic and acoustic influence mines. The MCM class also conducts coordinated operations with airborne and other mine countermeasure forces. This is the largest US Navy minesweeper to date and the first MCM ships to be built in America in nearly 30 years. The last three MCM ships were purchased in 1990, bringing the total to 14 fully deployable, oceangoing Avenger class ships. These ships use sonar and video systems, cable cutters and a mine detonating device that can be released and detonated by remote control. The ships are constructed of wood covered with glass-reinforced plastic (GRP) sheathing. The *Sentry* is equipped with the AN/SLQ-37(v) Standard Magnetic/Acoustic Influence Minesweeping System. It consists of a straight tail magnetic sweep (M Mk 5A) combined with the A Mk 4(v) and/or A Mk 6(b) acoustic sweeps. The system can be configured several ways, including diverting the magnetic cable and/or the acoustic devices by using components of the AN/SLQ-38 mechanical sweep gear.

SPECIFICATIONS

Type:	minehunter
Crew:	84
Displacement:	1382 tonnes (1360 tons)
Length:	68.3m (224ft)
Beam:	11.9m (39ft)
Draught:	3.4m (11.75ft)
Speed:	13.5 knots
Range:	unknown
Missiles:	none
Guns:	2 x 12.7mm
Torpedoes:	none
Air Search Radar:	none
Surface Radar:	SC Cardion SPS 55
Navigation:	SSN-2V Precise Integrated Navigation System (PINS)
Fire Control Radar:	none
Propulsion:	4 x diesels

USS SPRUANCE

USS *Spruance* is the first destroyer to be back-fitted with the MK 41 Vertical Launching System (VLS). This allows her to engage shore-based and naval surface targets at long range. In addition, state-of-the-art computer and satellite technology allow the ships of this class to launch up to 61 precision-guided Tomahawk cruise missiles at land targets as far away as 1120km (700 miles). For example, ships of this class fired a total of 112 Tomahawk cruise missiles into Iraq during Operation Desert Storm in 1991 against important targets. They have subsequently been used for pre-emptive strikes at the direction of National Command Authorities against both Iraq and Bosnia. These ships have traditionally had a major role in Naval Surface Fire Support for troops ashore, employing Harpoon anti-ship missiles and two 127mm (5in) guns (also used for air defence and shore bombardment). The Harpoon missile system is proven effective in engaging shipping at intermediate ranges. The two MK 45 lightweight 5in 54-calibre guns can throw a projectile over 19.2km (12 miles) with a firing rate of 20 rounds per minute. The 5in gun represents a major step forward in medium-calibre ordnance for the US Navy, and the result is a weapon which allows a single man in a control centre to fire a barrage of 20 shells without assistance.

SPECIFICATIONS

Type:	destroyer
Crew:	382
Displacement:	9144 tonnes (9000 tons)
Length:	171.7m (563.32ft)
Beam:	16.8m (55.11ft)
Draught:	8.8m (29ft)
Speed:	33 knots
Range:	9600km (6000 miles)
Missiles:	SLCM, Harpoon, Sea Sparrow
Guns:	2 x 127mm, 2 x Phalanx
Torpedoes:	Mk 46
Helicopters:	2 x SH-60B LAMPS III
Aircraft:	none
Air Search Radar:	SPS-40E
Surface Radar:	SPS-55
Fire Control Radar:	SPG-60, SPQ-9A
Propulsion:	4 x gas turbines, 80,000shp

USS STOUT

This Arleigh Burke class destroyer was commissioned on 13 August 1994. The design process for ships built at Ingalls Shipbuilding is accomplished using a three-dimensional computer-aided design (CAD) system, which is linked with an integrated computer-aided manufacturing (CAM) production network of host-based computers and localized minicomputers throughout the shipyard. The technology significantly enhances design efficiency, and reduces the number of manual steps involved in converting design drawings to ship components, improving productivity and efficiency. During the construction of a DDG-51 destroyer, hundreds of sub-assemblies are built and outfitted with piping sections, ventilation ducting and other shipboard hardware. These sub-assemblies are joined to form dozens of assemblies, which are then joined to form the ship's hull. During the assembly integration process, the ship is outfitted with larger equipment items, such as electrical panels, propulsion equipment and generators. The ship's superstructures are lifted atop the ship's midsection early in the assembly process, facilitating the early activation of electrical and electronic equipment. When hull integration is complete, the ship is moved over land via a wheel-on-rail transfer system, and onto the shipyard's launch and recovery drydock.

SPECIFICATIONS

Type:	guided missile destroyer
Crew:	323
Displacement:	8433 tonnes (8300 tons)
Length:	153.8m (504.59ft)
Beam:	20.4m (66.92ft)
Draught:	9.44m (31ft)
Speed:	32 knots
Range:	7040km (4400 miles)
Missiles:	SLCM, Harpoon ASM, SM-2 SAM
Guns:	1 x 127mm, 2 x Phalanx, 2 x 25mm
Torpedoes:	Mk 46, Mk 50
Helicopters:	none
Aircraft:	none
Air Search Radar:	SPY-1D 3D
Surface Radar:	SPS-67(V)3
Fire Control Radar:	3 x SPG-62
Propulsion:	4 x gas turbines, 100,000shp

USS TARAWA

The Tarawa class can fulfil a number of roles: flagship for embarked amphibious squadron, flag or general officer staff; aircraft carrier; amphibious assault launching platform, employing a variety of surface assault craft including the navy's Landing Craft Utility (LCU), and other amphibious assault vehicles; hospital ship with 17 intensive care beds, four operating rooms, 300 beds, a 1000-unit blood bank, full dental facilities and orthopedics, trauma, general surgery, and x-ray capabilities; command and control ship; and assault provisions carrier able to sustain embarked forces with fuel, ammunition and other supplies. The LHA's flight deck can handle up to 10 helicopters simultaneously, as well as the AV-8B Harrier aircraft. There is also a large well deck in the stem of the ship for a number of amphibious assault craft, both displacement hull and air cushion. The ships have an extensive command, communication and control suite. These electronic systems give the amphibious task force commander nearly unlimited versatility in directing the assault mission. The heart of the LHA's electronic system is a tactical amphibious warfare computer which keeps track of the landing force's positions after leaving the ship, tracks enemy targets ashore and directs the targeting of the ship's guns and missiles.

SPECIFICATIONS

Type:	amphibious assault ship
Crew:	964 plus 1900 US Marines
Displacement:	40,564 tonnes (39,925 tons)
Length:	254.2m (833ft)
Beam:	40.2m (131ft)
Draught:	7.9m (25.91ft)
Speed:	24 knots
Range:	16,000km (10,000 miles)
Missiles:	RAM
Guns:	2 x Phalanx
Torpedoes:	none
Helicopters:	9 x CH-53D, 12 x CH-46D/E
Aircraft:	6 x AV-8B
Air Search Radar:	SPS-40, SPS-48E
Surface Radar:	SPS-67
Fire Control Radar:	Mk 23 TAS
Propulsion:	2 x steam turbines, 70,000shp

USS TICONDEROGA

The lead ship of the Ticonderoga class of guided missile cruisers, she is equipped with two SH-2G Seasprite helicopters. The SH-2G Super Seasprite was originally developed in the mid-1950s as a shipboard utility helicopter. Utilizing a unique blade flap design on the main rotors, the aerodynamic action of the flaps allows the pilot to fly without the aid of hydraulic assistance. The SH-2G is configured specifically to respond to the Light Airborne Multi-Purpose System (LAMPS) requirement of the United States Navy. The LAMPS concept extends the search and attack capabilities of carrier and convoy escort vessels over the horizon through the use of radar/electronic support measures (ESM) equipped helicopters. Primary missions of the SH-2G are anti-submarine warfare (ASW) and anti-ship surveillance and targeting (ASST). Secondary missions include search and rescue, vertical replenishment, medical evacuation, communications relay, personnel transfer, surveillance and reconnaissance, post-attack damage assessment, and naval gunfire spotting. Armament systems consist of two search stores systems (sonobuoys and marine location markers), an external weapons/stores system for external fuel tanks or torpedoes, and a countermeasures dispensing system. The other ships in the class carry two SH-60B Seahawk helicopters.

SPECIFICATIONS

Type:	guided missile cruiser
Crew:	344
Displacement:	10,116 tonnes (9957 tons)
Length:	172.8m (566.92ft)
Beam:	16.8m (55.11ft)
Draught:	9.5m (31.16ft)
Speed:	30 knots
Range:	9600km (6000 miles)
Missiles:	SLCM, Harpoon, SM-2MR M
Guns:	2 x 127mm, 2 x Phalanx, 2 x 25mm
Torpedoes:	Mk 46, Mk 50
Helicopters:	2 x SH-2G
Aircraft:	none
Air Search Radar:	SPS-49
Surface Radar:	SPS-55
Fire Control Radar:	SPQ-9
Propulsion:	4 x gas turbines, 80,000shp

USS TRENTON

The Austin class of amphibious ships is configured as flagships to provide extensive command, control and communications facilities to support amphibious landings. In an amphibious assault, the ship would normally function as the Primary Control Ship that would be responsible for coordinating boat waves and vectoring landing craft to the beach. A secondary mission is evacuation and civilian disaster relief. Hundreds of tons of relief materials can be carried aboard and delivered to disaster victims within minutes of the ship's arrival on the scene. Her medical and dental facilities can provide limited hospitalization care, as well as out-patient treatment for hundreds of sick or injured. The ship has a large flight deck for helicopter operations and a well deck that carries amphibious landing vehicles. The ships can carry one LCAC, or one utility landing craft (LCU) boat, or four mechanized landing craft (LCM), and six CH-46D/E helicopters, or three CH-53D helicopters, and 900 troops. To facilitate the docking and loading of various-sized landing craft, the ship can ballast down in the water, thereby flooding the well deck to enable the landing craft to enter the well deck through the stern gate door. Once docked inside the well deck, troops, supplies and combat equipment can be loaded into amphibious boats and vehicles.

SPECIFICATIONS

Type:	amphibious assault ship
Crew:	420 plus 900 US Marines
Displacement:	17,520 tonnes (17,244 tons)
Length:	173.8m (570.2ft)
Beam:	25.6m (84ft)
Draught:	7.2m (23.62ft)
Speed:	21 knots
Range:	12,320km (7700 miles)
Missiles:	none
Guns:	2 x Phalanx, 2 x 25mm
Torpedoes:	none
Helicopters:	6 x CH-46D/E or 3 x CH-53D
Aircraft:	none
Air Search Radar:	SPS-40 B/C
Surface Radar:	SPS-60
Fire Control Radar:	none
Propulsion:	2 x steam turbines, 24,000shp

USS WASP

These ships provide the means to deliver, command and support all elements of a Marine Landing Force in an assault by air and amphibious craft. In carrying out their mission, the ships have the option of utilizing various combinations of helicopters, Harrier II (AV-8B) close air support jump jets and air cushion landing craft (LCAC), as well as conventional landing craft and assault vehicles. Off the landing beach, the ship can ballast more than 15,240 tonnes (15,000 tons) of sea water for trimming during landing craft launch and recovery operations in the well deck. Wasp class ships can also provide command and control and aircraft facilities for sea control missions, while operating with an Aircraft Carrier Battle Group. They transport and land ashore not only troops, but also tanks, trucks, jeeps, other vehicles, artillery, ammunition and various supplies necessary to support the amphibious assault mission. Air cushion landing craft can "fly" out of the dry well deck, or the well deck can be ballasted down for conventional craft to float out. Helicopter flights also transfer troops and equipment to the beach, while the ship's air traffic control capability simultaneously directs close air tactical support provided by embarked jet aircraft and helicopter gunships. The two steam propulsion plants are the largest in the US Navy.

SPECIFICATIONS

Type:	amphibious ship
Crew:	1108 plus 1894 US Marines
Displacement:	41,180 tonnes (40,532 tons)
Length:	257.3m (844.16ft)
Beam:	42.7m (140ft)
Draught:	8.1m (26.57ft)
Speed:	22 knots
Range:	15,200km (9500 miles)
Missiles:	Sea Sparrow SAM, RAM
Guns:	2 x Phalanx, 3 x 25mm
Torpedoes:	none
Helicopters:	42 x CH-46E Sea Knight
Aircraft:	6–8 x AV-8B Harrier
Air Search Radar:	SPS-48E, SPS-49(V)7
Surface Radar:	SPS-67
Fire Control Radar:	Mk 91
Propulsion:	2 x steam turbines, 70,000shp

INDEX

Admiral Kuznetsov 148
Admiral Panteleyev 146
air defence destroyers 102
air defence frigates 145
aircraft carriers 100, 106, 108, 122, 123, 136, 161, 163, 168, 172, 176
AIP (see Air Independent Propulsion)
ammunition ships 171
amphibious assault ships 170, 185, 187
amphibious command ships 155
amphibious helicopter carriers 129
amphibious operations 35, 77
amphibious ships 156, 188
Andrea Doria 144
anti-shipping 16, 21, 46, 65, 68
anti-submarine 16, 32, 38, 46, 68, 81, 89
anti-surface 17, 32
USS *Antietam* 152
HMAS *Anzac* 99
HMS *Ark Royal* 122, 123
USS *Arleigh Burke* 153
Arleigh Burke class 153, 154, 177, 180, 184
attack submarines
  Agosta class 38, 39
  Akula class 48, 54
  Akula II class 43
  Améthyste class 16
  Astute class 73
  Collins class 9
  Han (Type 091) 9
  Improved Akula class 59
  Improved Los Angeles class 83, 93
  Kilo class 9, 13, 26, 46, 60
  Los Angeles class 84, 89
  Oscar II class 44, 52
  Salvatore Todaro class 28
  Scorpène class 64
  Seawolf class 85, 86
  Severodvinsk class 56
  Sierra I class 49
  Sierra II class 55
  Sindhugosh class 26
  Sturgeon class 88, 92
  Swiftsure class 74, 75
  Trafalgar class 76, 77
  Tupi class 10
  Type 209 10, 24, 25, 62
  Type 212A 23, 28
  Victor III class 45
  Yushio class 31
Austin class 187
Australia 99
Avenger class 182

Baleares 150
ballistic missile submarines
  Borei class 61
  Delta III class 58
  Delta IV class 47, 53
  L'Inflexible class 18, 19
  Le Triomphant class 20, 21
  Ohio class 90, 91, 95
  Typhoon class 40, 57
  Vanguard class 78, 79
  Xia (Type 092) 12
USS *Barry* 154
Black Hole 26
USS *Blue Ridge* 155
USS *Boxer* 156
Brandenburg 115
Brazil 100
Bremen 116
USS *Bridge* 157

HMS *Campbeltown* 118, 120
Canada 101-103
USS *Cape St George* 158
Casiopee 105
Cassard 104
HMS *Cattistock* 119
Centaur class 135
Charles De Gaulle 106, 108
Charles F. Adams class 117
Clemenceau class 108
Colossus 100
combat stores ship 175
Concord 175

USS *Constellation* 159
HMS *Cornwall* 120
corvettes 141
covert infiltration 34, 35, 36, 88
  covert operations 80, 82
  covert seafloor operations 41
cruise missile 44, 48, 56

DDS (Dry Dock Shelter) 87, 88, 92
destroyers 104, 114, 117, 121, 124, 125, 134, 139, 162, 183
disaster response 37
dock landing ship 165, 166
Duke class 118, 130, 133, 126-8, 133
USS *Dwight D. Eisenhower* 160

HMS *Edinburgh* 121, 125
electronic systems
  Air Independent Propulsion 12, 14, 23, 28, 33, 39, 64, 67
  AN/BPS-15 radar 83, 89, 93
  AN/BQQ-5D sonar 84
  Atlas Electronik CSU 90-2 sonar 68
  BAe 2054 sonar 78, 79
  BAe Type 2007 sonar 76, 77
  BPS-14/15 radar 88, 92
  BPS-15A radar 87
  BPS-16 radar 85, 86, 94, 96, 97
  BQS-13 sonar 88, 90, 92, 95
  BSY-1 sonar 83, 89, 93
  BSY-2 sonar 85, 86, 94
  IBM BQQ-6 sonar 87, 91
  Irtysh-Amfora sonar 50, 56
  ISUS 90-20 sonar suite 27
  Kelvin Hughes Type 1007 64
  Krupp-Atlas Electronik CSU-83 sonar 8, 14, 37, 62, 66, 67, 72
  Marconi 2074 sonar 74, 75
  Medvyedista-671 45
  Medvyedista-945 43, 49, 54, 59
  Medvyedista-971 56

MGK-400 sonar 26
MGK-540 sonar 54, 61
Omnibus-BDRM 47
Rubikon sonar 46, 58, 60
Shark Gill sonar 40, 44, 57
Shlyuz navigation system 47, 53
Signaal SIASS-Z sonar 70
Skat sonar 47, 53, 55, 59
SUBTICS combat system 64
Terma radar 66, 67, 68, 69
Thales 2076 sonar 73
Thales Calypso sonar 10
Thomson CSF Calypso II 20, 24, 25
Thomson Sintra DSUX 23 18, 19
Thomson Sintra Scylla sonar 9
Thomson Sintra TSM 2272 32
Thomson-CSF DRUA 35
Tobol navigation system 51, 58
ZQQ-4 sonar 31
ZQQ-5B sonar 29, 30
USS *Enterprise* 161
Eridan class 105
exfiltration 35

Falklands War 16, 75
  FOST (French Strategic Oceanic Force) 19, 20
fast combat support ships 157
USS *Fife* 162
fleet oiler 179
Floreal 107
Foch 108
Foudre 109
France 104-114
frigates 99, 110, 112, 115, 116, 118, 120, 126, 127, 128, 133, 138, 140, 142, 143, 149, 150
MM *Francesco Mimbelli* 139

USS *George Washington* 163
Georges Leygues 110
Georges Leygues class 110, 112
Germany 115-117

Giuseppe Garibaldi 136, 144
Great Britain 118-134
guided missile cruisers 152, 158, 173, 191, 196
guided missile destroyers 103, 146, 147, 153, 154, 177, 180, 184
guided missile frigates 166, 167, 170
USS *Gunston Hall* 164

Halifax 101
USS *Harpers Ferry* 165
helicopter carriers 111
helicopter cruisers 144
HLES 80 steel 39
Hunt class 119
hunter-killer 69, 73, 75, 76, 77
HY130 steel 20

HMS *Illustrious* 122
India 135
USS *Ingraham* 166
intelligence gathering 88
HMS *Invincible* 122, 123
Invincible class 122, 123
Iroquois 102
Italy 136-144

Jacob Van Heemskerck 145
USS *Jarrett* 167
Jeanne d'Arc 111
USS *John C. Stennis* 168
USS *John F. Kennedy* 169

USS *Kauffman* 170
USNS *Kilauea* 171
USS *Kitty Hawk* 172
Knox class 150
Kortenaer class 145
Kosovo crisis 75, 78

Krivak I frigate 149
Kuznetsov 148

La Motte-Picquet 112
USS *Lake Erie* 173
HMS *Lancaster* 123
landing platform docks 109
large landing ships 132
L'Audacieuse 113
Lerici 137
Libeccio 138
littoral operations 9, 15, 37, 60, 65, 66, 97, 78, 79
HMS *Liverpool* 124
Luda 103
MM *Luigi Durand De La Penne* 139
Lupo class 138, 142, 143
Lutjens 117

Maestrale 140
Maestrale class 138, 140
HMS *Manchester* 125
Marine Expeditionary Unit 156
HMS *Marlborough* 126
Mars class 175
midget submarine 35, 36
mine countermeasures vessels 119
minehunters 105, 130, 137, 178, 182
Minerva 141
MIRV (see multiple independently-targetable re-entry vehicle)
MRV (see multiple re-entry vehicle)
HMS *Montrose* 127
multiple independently-targetable re-entry vehicle 53, 57, 58, 78, 79, 90, 91
multiple re-entry vehicle 21
multirole frigates 101

NATO 99, 100, 101, 146
USS *Nassau* 174

INDEX 191

NAWCAD (see Naval Air Warfare Center Aircraft Division)
New Zealand 99
USNS *Niagra Falls* 175
USS *Nimitz* 176
Nimitz class 160, 163, 176
HMS *Norfolk* 128
NS110 steel 29
NS80 steel 31

HMS *Ocean* 129
ocean engineering 41
Ocean Interface 86
ocean survey ships 131
offshore patrol boats 113
Oliver Hazard Perry class 166, 167, 170
Operation Enduring Freedom 32
organizations
   Admiralty Shipyard 46
   BAe Systems Marine 73
   DCN International 16, 17, 18, 19, 20, 21, 38, 39, 64
   Droogdok Maatschappij B.V. 32, 33
   Electric Boat 81, 82, 85, 86, 87, 90, 91, 93, 94, 95, 97
   Ingalls Shipbuilding 92
   Japanese Maritime Self Defense Force 29, 30, 31
   Kockums 14, 66, 67, 68, 69
   Naval Air Warfare Center Aircraft Division 81
   Newport News 83, 84, 88, 89, 96, 97
   North Atlantic Treaty Organization 15, 20, 26, 37, 49
   Sevmash 48, 49, 50, 51, 52, 53, 58, 61
   Thyssen Nordseewerke 8
   United Nations 37
   US Special Operations Command 80
   Vickers Shipbuilding 74, 75, 76, 77
Orsa 142

USS *Oscar Austin* 177
USS *Osprey* 178

patrol frigates 107
patrol submarines
   Changbogo class 62
   Delfin class 63
   Glavkos class 24
   Gotland class 68
   Harushio class 29
   Hashmat class 38
   Khalid class 39
   Kronborg class 14
   Moray class 33
   Nacken class 66
   Oyashio class 30
   Primo Longobardo class 27
   Romeo class 34
   Rubis Améthyste class 17
   Santa Cruz class 8
   Shishumar class 25
   Sodermanland class 67
   Tumleren class 17
   Type 206A 22
   Ula class 37
   Vastergotland class 69
   Walrus 2 class 32
peace-enforcement 9, 15, 93
peacekeeping 9, 15, 67, 93
USS *Platte* 179
USS *Porter* 180
USS *Princeton* 181
Principe de Asturias 151
Puget Sound Naval Shipyard 161
Pytlivyy 149

reconnaissance 80
remotely piloted vehicle 131
research vessel 41, 42, 82
Russia 146-149

Sagittario 142, 143
SAM (surface-to-air missile) 26

San Jose 175
HMS *Sandown* 130
satellite communcations 23
HMS *Scott* 131
USS *Sentry* 182
SH-60B Seahawk 81
Sheffield class 124, 125, 134
Ship Submersible Ballistic Nuclear 12, 18, 19, 20, 21, 40, 47, 50, 51, 57, 58, 61, 78, 79, 87, 90, 91, 95
RFA *Sir Galahad* 132
Sir Lancelot class 132
Siroco 109
SNLE-NG 20
HMS *Somerset* 133
HMS *Southampton* 134
sound dampening 43, 55
sound-absorbent tiles 40, 57
Sovremenny 147
Spain 150-51
special forces 35, 36, 51, 80, 82, 86
special operations submarines
   Advanced SEAL delivery system 80, 86, 96, 97
   Benjamin Franklin class 87
   Paltus class 42
   NR-1 82
   Sang-O class 35
   Yugo class 36
   Yankee Stretch 51
USS *Spruance* 183
Spruance class 162, 183
SSBN (see Ship Submersible Ballistic Nuclear)
SSN (see attack submarines)
strategic deterrence 18, 19, 21
surveillance 32, 68, 77
USS *Stout* 184
Suffren 114

USS *Tarawa* 185
Tarawa class 174, 185
Tbilisi 148
thrusters 82
USS *Ticonderoga* 186

Ticonderoga class 152, 158, 173, 186
training/trials submarine
    Tang class 81
    Yankee Pod 50
USS *Trenton* 187
TRIAD 90, 95
Type 22 class 118, 120, 123
Type 23 Duke class 118, 120, 123, 126-27, 133
Type 42 destroyers 121, 124, 125, 134
Type 82 125
Type 123 Brandenburg class 115

Udaloy class 146
Uniform class 41
United States 152-188
US Navy SEALs 80, 87, 88
USSOCOM (see US Special Operations Command)

Viking project 14
INS *Viraat* 135
Vittorio Veneto 144
USS *Wasp* 188
Wasp class 156, 188
weapons
    ADCAP (Advanced Capability) torpedo 89, 97
    AEG SST-4 torpedo 72
    C-4 Trident I SLBM 90, 91
    D-5 Trident II SLBM 78, 79, 90, 93
    Exocet missile 16, 17, 18, 21, 39, 65
    Granat missile 48, 55, 56
    Klub missile 13
    M4 SLBM 18
    M45 SLBM 21
    mines 8, 12, 38, 36, 78, 80, 85, 91, 96
    Mk48 torpedo 83, 85, 89, 92
    Mk60 CAPTOR mine 97
    R-27 SLBM 50
    R-29 SLBM 47, 53

RSM-52 SLBM 57
SA-N-5 Strela 48
Spearfish torpedo 73, 76, 78, 79
SS-N-15 Starfish missile 48
SS-N-16 Stallion missile 48
SS-N-18 SLBM 58
SS-N-28 SLBM 40
SS-N-29 SLBM 61
Sub Harpoon missile 9, 24, 29, 30, 33, 71, 73, 74, 85, 97
Tigerfish torpedo 76, 78, 79
Tomahawk cruise missile 73, 75, 76, 77, 78, 79, 83, 83, 85, 86, 89, 92, 93, 94, 96, 97
Torpedo 2000 68
Type 53 wake-homing torpedo 26
Whidbey Island class 164

YJ-82 missile 11

Picture Credits

All photographs The Robert Hunt Library except the following:

Bae Systems: 79, 119, 131
Chinese Defence Today (www.sinodefence.com): 11, 12
Electric Boat: 94
Jane's Information Group: 99, 100, 101, 103, 104, 106, 107, 109, 115, 116, 117, 120, 124, 129, 136, 146, 149, 150, 157, 159, 160, 162, 163, 165, 167, 168, 176, 180, 185
Kockums: 67, 68, 69
MPL: 11, 17, 38
Nordseewerke GmbH: 37
PA Photos: 26
Private Collection: 8, 9, 10, 14, 15, 16, 17, 18, 20, 21, 22, 23, 24, 26, 27, 28, 29, 30, 31, 32, 33, 34, 35, 36, 38, 39, 41, 42, 43, 44, 46, 48, 49, 50, 51, 53, 54, 55, 56, 57, 58, 60, 61, 62, 63, 64, 65, 66, 70, 71, 72, 74, 75, 76, 77, 80, 81, 86, 87, 88, 89, 92, 93, 96, 97
Rex: 44
TRH: 25, 45, 47, 52, 59, 78, 79, 102, 105, 108, 110, 111, 112, 113, 114, 118, 121, 122, 123, 125, 127, 128, 129, 131, 133, 134, 135, 137, 138, 139, 140, 141, 142, 143, 144, 145, 147, 148, 151
US DoD: 82, 83, 84, 85, 90, 91, 95, 152, 153, 154, 155, 156, 158, 161, 164, 166, 168, 169, 170, 171, 172, 173, 174, 175, 177, 178, 179, 181, 182, 183, 184, 186, 187